ENIGMA OF THE SKIES

Unveiling the Secrets of Auroras

ENIGMA OF THE SKIES
Unveiling the Secrets of Auroras

Yohsuke Kamide
Rikubetsu Space and Earth Science Museum, Japan &
Nagoya University, Japan

Yoshi Otsuka
Nanook Aurora Tours, Canada

edited by
Yusuke Ebihara
Research Institute for Sustainable Humanosphere (RISH),
Kyoto University, Japan

World Scientific

NEW JERSEY · LONDON · SINGAPORE · BEIJING · SHANGHAI · HONG KONG · TAIPEI · CHENNAI · TOKYO

Published by

World Scientific Publishing Co. Pte. Ltd.

5 Toh Tuck Link, Singapore 596224

USA office: 27 Warren Street, Suite 401-402, Hackensack, NJ 07601

UK office: 57 Shelton Street, Covent Garden, London WC2H 9HE

Library of Congress Cataloging-in-Publication Data

Names: Kamide, Y. (Yōsuke), 1943-2021, author. | Otsuka, Yoshi
 (Photographer), photographer. | Ebihara, Yusuke, editor.
Title: Enigma of the skies : unveiling the secrets of auroras / Yohsuke
 Kamide, Rikubetsu Space and Earth Science Museum, Japan & Nagoya
 University, Japan, Yoshi Otsuka, Nanook Aurora Guide, Canada ; edited by
 Yusuke Ebihara, Research Institute for Sustainable Humanosphere (RISH),
 Kyoto University, Japan.
Description: New Jersey : World Scientific, [2023] | Includes
 bibliographical references. |
Identifiers: LCCN 2022037210 | ISBN 9789811228773 (hardcover) |
 ISBN 9789811230394 (paperback) | ISBN 9789811228780 (ebook for institutions)
 | ISBN 9789811228797 (ebook for individuals)
Subjects: LCSH: Auroras. | Auroras--Pictorial works.
Classification: LCC QC971 .K348 2023 | DDC 538/.768--dc23/eng20221024
LC record available at https://lccn.loc.gov/2022037210

British Library Cataloguing-in-Publication Data
A catalogue record for this book is available from the British Library.

October 2023 Edition

For any available supplementary material, please visit
https://www.worldscientific.com/worldscibooks/10.1142/12057#t=suppl

Desk Editor: Amanda Yun
Design and layout: Lionel Seow

Foreword

Prof. Yohsuke Kamide, widely known in the scientific world simply as "Kamide" or "Kamide Sensei", delighted in explaining to the public the wonders of space physics and the way the solar wind influences Earth's magnetosphere and upper atmosphere. This book is an engaging product of that zeal. Sadly, it is his last gift to us before his death in 2021.

Kamide and I met in Boulder as visiting scientists at the National Geophysical Data Center, which operated the World Data Center A for geomagnetism and solar-terrestrial data. We analyzed geomagnetic variations from magnetometers all over the Northern Hemisphere to look at patterns of ionospheric electric fields and currents. This collaboration continued over our careers and grew to include many other colleagues interested in this topic.

Kamide was always getting things done, whether in scientific research, in education, in organization of international scientific efforts, or in everyday life. His prolific scientific publications cover a wide range of topics involving the science behind the aurora. He worked in several national and international scientific societies to arrange conferences, publish scientific journals and books, and create new organizations like the Asia Oceania Geosciences Society. One of his proudest achievements was helping bring the quadrennial meeting of the International Union of Geodesy and Geophysics to Sapporo in 2003 and arranging for the Japanese Emperor, a trained scientist, to speak there.

On one extended working visit to Boulder, Kamide brought his family along. He found a suitable house and he adroitly negotiated the purchase of a used car, which he sold for about the same price at the end of his stay. Since our children were of similar ages, our families got together for barbeques and yard games. Years later, one of his daughters stayed with us after high-school graduation.

Kamide encouraged me to visit him in Japan, both to continue our joint work and to give me insight into Japanese life. He helped arrange my visit to Kyoto Sangyo University, and we traveled to meet colleagues in Tokyo and

Fukuoka. It was a beautiful autumn, with the maple trees in full color. Kamide liked to show me the modern side of Japan as well as its history and some typically Japanese cultural features. For our travels, we rode the precision-timed Shinkansen trains—a trademark of Japanese efficiency. He pointed out that it is safe to walk everywhere in Japan, and I took advantage of that to explore. The university campus was lively, with noontime martial arts groups chanting through their routines. Kamide felt it was important for me to have a 10 minute meeting with the university president, and informed me how to dress properly with coat and tie and how to greet the president with a respectful bow.

Kamide's interest in bringing solar-terrestrial science to the broad public was a lifelong passion. As an early effort, he arranged with the publisher of the Japanese edition of the *Scientific American* magazine to publish a special issue exclusively on solar-terrestrial topics. He enlisted experts from around the world to write articles for it. I marveled to learn how he was invited to talk about auroras at a union meeting of Japanese taxi drivers. He organized a series of comic books on solar-terrestrial topics to attract the interest of children (and adults); these are available through the Scientific Committee on Solar-Terrestrial Physics. Kamide also worked on many publications like the present book, aimed at the broad public.

Kamide was a friend. We helped each other on scientific and personal projects and traded ideas on many topics. We deeply respected each other's particular strengths. He often asked for my advice on English usage in an international context, such as the best wording and punctuation choices for the title of the journal that became *Earth, Planets and Space*. I last saw him in 2017 at a conference in Tokyo. Despite illness, he made a special effort to meet me at a restaurant for dinner, where we recounted old times. I was greatly saddened by his death and the end of his contributions to science and education, contributions that will have lasting impact.

Arthur D. Richmond
Senior Scientist Emeritus
High Altitude Observatory
National Center for Atmospheric Research

Preface

This book represents a joint endeavor by two people, a scientist in space physics, aurora generation and mechanisms, and a photographer of natural phenomena, primarily aurora borealis. Our experiences and techniques differ, but our common ground is the quest to unveil the aurora process and its mysterious beauty.

Auroras are one of the most beautiful and hard to catch phenomena that human beings can experience. Many people throughout history, from all walks of life, have participated in the quest to understand auroras and our natural world. There has been progress, but we are still far from understanding auroras' whole structure.

What we know is that auroras are created through a complicated, intricate relationship between the Sun and Earth. We also know its nature can sustain or disrupt our lives here on Earth. While we each have different experiences and techniques for understanding the Earth's system, there is, at least, one point in common.

We must look at the entire story to really understand. Appreciating everything that relates to the occurrence of auroras (history, art, writings and scientific interpretation) is crucial to understanding the natural world and our place in it.

The Sun is the birth-place of auroras. To the residents of Earth, the existence of the Sun has been absolute. To some, the Sun is seen as a god and a mother. However, it has recently been known that the Sun is rough, wild and dynamic—completely different from the image of a nurturing God or mother. The consequence of its rough interaction with the Earth are the beautiful auroras we see in the northern and southern hemispheres.

Anyone encountering this dynamically dancing light show in the polar sky may feel that only a god could paint the sky that way. Each of us tries to express this mysterious light as accurately as possible, even though the methods and languages we use are completely different. Each human perspective is another part of this intricate natural dance.

Acknowledgments

We would like to thank our copy editors Laura Love and Gregory Lee, and our editor Amanda Yun for the many improvements they have made to this book. We would also like to thank Lionel Seow for his input and work on the layout of this book.

Yohsuke Kamide
Director, Rikubetsu Space and Earth Science Museum, Japan & Professor Emeritus at Nagoya University, Japan

Yoshi Otsuka
President, Nanook Aurora Tours, Yellowknife, NT, Canada

The editor, **Yusuke Ebihara**, would like to express his deepest gratitude to the following people for their support.

Dr. Hisashi Hayakawa, *Nagoya University, Japan*

Prof. Akira Kadokura, *National Institute of Polar Research, Japan*

Prof. Yoshihiro Kakinami, *Hokkaido Information University, Japan*

Dr. Satoshi Masuda, *Nagoya University, Japan*

Prof. Ayako Matsuoka, *Kyoto University, Japan*

Ms. Yoko Odagi, *Kyoto University, Japan*

Dr. Masahiro Terada, *Kyoto University, Japan*

Dr. Tatsuhiro Yokoyama, *Kyoto University, Japan*

About the Authors

Yohsuke Kamide was Professor Emeritus of Nagoya University and Director of the Rikubetsu Space and Earth Museum. He obtained his PhD from the University of Tokyo, majoring in space science, particularly in solar-terrestrial physics. Prof Kamide also previously held research and teaching positions in Hokkaido University, University of Tokyo, Kyoto University, University of Alaska, and University of Colorado, in addition to positions in USA's National Oceanic and Atmospheric Administration (NOAA), National Center for Atmospheric Research (NCAR), and National Geophysical Data Center (NGDC), Space Environment Laboratory (SEL).

Prof. Kamide published ~350 scientific papers and ~30 books, the most prominent one being the *Handbook of the Solar Terrestrial Environment* (Springer). He was Vice President of the Local Organizing Committee (LOC) of IUGG 2003 in Sapporo, and was responsible for inviting the Japanese Emperor and Empress. He had also served as editor of the *Journal of Geophysical Research: Space Physics* for 11 years, and is one of the three co-founders of the Asia Oceania Geosciences Society (AOGS).

Prof. Kamide was a fellow of Royal Astronomical Society (RAS), UK and Japan Geoscience Union (JpGU), and had been given awards by RAS, including its prestigious Price Medal, and the American Geophysical Union (AGU). He had also been awarded the Hasegawa-Nagata Medal, and AOGS's Axford Medal. To honor his scientific contributions, the Kamide Lecture was instituted at AOGS.

Yoshi Otsuka graduated from Nakanoshima Art School and worked as a graphic designer in Japan. He has been an aurora guide and photographer in Yellowknife, Canada since 2001.

About the Editor

Yusuke Ebihara is an Associate Professor of Kyoto University, Japan. He received his Ph.D. from the Graduate University for Advanced Studies, Japan. He is the author or coauthor of more than 200 scientific publications and 6 books. He received the Tanakadate Award of the Society of Geomagnetism and Earth, Planetary and Space Sciences (SGEPSS) in 2012, and Nishida Prize of the Japan Geoscience Union (JpGU) in 2017. He served as Associate Editor of the *Journal of Geophysical Research — Space Physics.*

Table of Contents

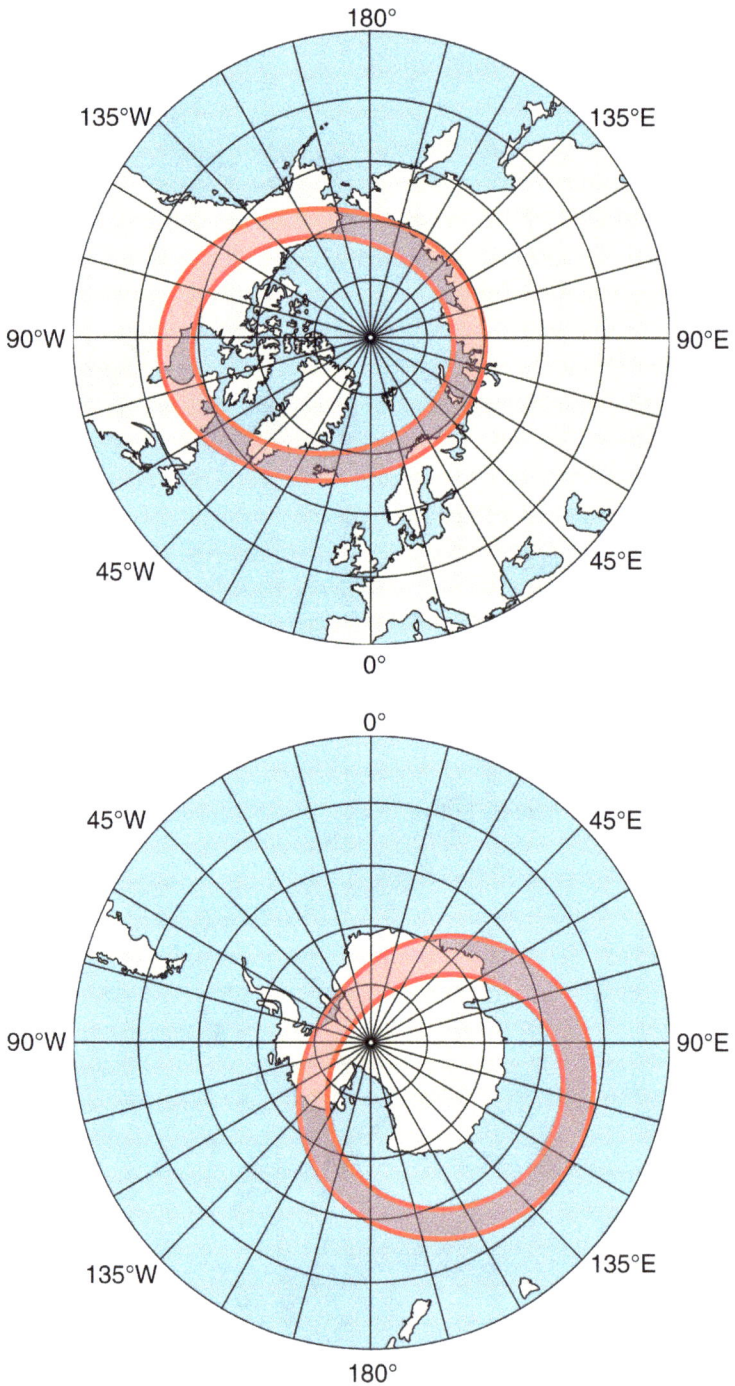

▲ Fig. 1-1. Auroral belts in the northern (top) and southern (bottom) hemispheres.

Chapter 1

Auroras Around the World

The aurora is a beautiful dancing manifestation of the delicate relationship between the Sun and the Earth. Over hundreds of years, many have been impressed by this beautiful and mysterious natural phenomenon. At the same time, their sense of curiosity has been piqued by how the Sun manages to continuously send messages to the Earth's surface, even in the form of auroral displays at night.

Throughout history, the Sun's messages have been revered and feared around the world. The word "aurora" originated in ancient Rome, where Aurora, the Goddess of Dawn, delivered sunshine every morning.

During ancient times, people postulated that auroras were generated far north of the northern horizon, near the end of this world, where their gods lived. They thought that auroras were signs of anger or disapproval from their gods.

Today, people regard auroras as "beautiful lights blasting out from the sky". Professional as well as amateur photographers, who are impressed and trapped by the power this mysterious light has, even go to the extent of "chasing" auroras around the world.

To which parts of the world should we travel to enjoy these mysterious auroras? And where can we see their dynamic but mysterious motions on a daily basis? There is a mistaken conception that auroras are more visible the closer one approaches the poles. Actually, the place to view auroras is a belt encircling the geomagnetic poles of the Earth. This is called an auroral belt; see Chapter 4 for the non-technical scientific explanation.

In the northern hemisphere, this belt makes a complete circle over Alaska, northern Canada, the southern part of Greenland, Iceland, Scandinavia, and along the northern edge of Siberia. In the southern hemisphere, an auroral belt of a similar size, as large as the Antarctic continent, exists: see Fig. 1-1.

1.1. Viewing Auroras from the Northern Hemisphere

In this chapter, we show a large number of selected beautiful auroral photos taken at many different places along the auroral belt. Though the mechanism behind the aurora itself is the same wherever we view them from, the different environments and scenery over which they occur give us many different manifestations of auroras.

From Alaska:

A large number of tourists visit Alaska every year to see auroras. Many of them are from Japan because of its relatively short travel distance to Fairbanks, the center of Alaska, which is directly under the auroral belt. One can see gorgeous auroral displays from the hotels in downtown Fairbanks.

Fairbanks is also where the main campus of the University of Alaska is situated. The University hosts one of the centers of auroral research, the Geophysical Institute at Poker Flat Research Range. Many visitors specially come here to take part in joint projects on auroras.

Also along the northern edge of Alaska, Utqiagvik (Barrow) is another popular place for tourists to view auroras. If one is lucky, auroras can also be seen over Anchorage, the biggest city in Alaska.

Figures 1-2 through 1-6 showcase some of the active auroras taken in the Alaska range by professional photographers.

◀ Fig. 1-2. Waiting for a brutal storm to pass in the Alaska Range paid off for photographer Norio Matsumoto. He camped all by himself on an unnamed small glacier, which was a tributary of the larger Ruth Glacier, 4,000 feet up in the Alaska Range, for two months. He built a snow cave to sleep in and ate rice, beans, and nori seaweed for his entire stay. His goal was to capture the "perfect" photographs from Denali. He attained his goal, as you can see here. Photo by Norio Matsumoto.

▲ Fig. 1-3. A beautiful half-moon rose at the Alaska Range on the bone-chilling December night this photo was taken. The moon lit up the entire landscape including Denali (20,310 feet) and the Ruth Glacier. It resulted in providing the perfect foreground for this photo. Such colorful northern lights have rarely been observed in the region, especially in the winter months, during which the weather is very unstable. Photo by Norio Matsumoto.

◀ Fig. 1-4. Norio Matsumoto captured this shot while camping alone in the middle of winter on a glacier in the Alaska Range. Solo camping in the Range is rare. There are usually only one or two people at a time in the Alaska Range. Norio has been spending two months there every winter since 2000. Although auroras are frequently present, it is rare to see them in the Alaska Range due to adverse weather. Photo by Norio Matsumoto.

▲ Fig. 1-5. Alone on a glacier with a front-seat view of the highest peak in Denali in late December, Norio Matsumoto attended to the familiar daily routine of his two-month-long winter camping trips in search of auroras. One frigid midnight, they appeared: a river of pale blue light filling a star-speckled sky behind the jagged peaks of the Alaska Range. The temperature was −40°C when he took this picture. Photo by Norio Matsumoto.

▼ Fig. 1-6. When sudden intensifications of auroras are exceptionally strong, the breakups can be seen even in downtown Fairbanks, Alaska. This photo shows a breakup as seen from a riverbank in downtown Fairbanks. Photo by Toshio Ushiyama.

From Canada:

The auroral belt in Canada runs nearly in an East–West direction, from the Hudson Bay to its border with Alaska. Anywhere along this belt is good for aurora viewing. However, there are only a few accessible locations in this belt, unless one is a professional mountaineer.

In terms of accessibility, Yellowknife, the capital of the Northwest Territories, (NWT) is recommended for its dry weather during winter, which attracts many tourists who want to connect with nature and enjoy winter activities while chasing auroras. It has a population of approximately 20,000 people and is located at the northern edge of the Great Slave Lake, the tenth-largest lake in the world.

Whitehorse is another attractive aurora destination, although the weather there is unpredictable due to its mountainous terrain. It is the capital of the Yukon Territories and offers a variety of other activities such as dogsledding, ice fishing, hot springs and winter festivals.

If one hikes along the western side of the Hudson Bay to Churchill, Manitoba, wild animal sightings and auroras are common.

Figures 1-7 through 1-12 show aurora photographs taken in and near Yellowknife. At this longitude, the geomagnetic latitude is approximately 10 degrees higher than the area's geographic latitude; because while the geomagnetic latitude of Yellowknife is 69 degrees, its geographic latitude is only 60 degrees.

◀ Fig. 1-7. Multiple breakups, taken at an aurora viewing tourist spot near Yellowknife NWT, Canada. Tourists stay inside a structure called a tipi until auroral movement is observed. Tipis were traditionally used by many aboriginal peoples in North America. Photo by You Koyanagi.

▲ Fig. 1-8. Aurora brightenings occur at at more than one place. It is not clear whether these multiple brightenings are related to each other. Photo by Sei Iwaihara, courtesy of Aurora Village, Yellowknife, Canada.

▼ Fig. 1-9. A rare red aurora appearing in downtown Yellowknife, Canada. Photo by Yoshi Otsuka.

▲ Fig. 1-10. Daylight hours in late-August in Yellowknife are long; it is still quite bright outside around 11 p.m. Even so, an intense aurora appears and can be observed from inside the town. This picture was taken from the parking lot of The Explorer Hotel, which stands on a small hill looking down on the town. Photo by Yoshi Otsuka.

▼ Fig. 1-11. Canadians often ask why Japanese people travel all the way to high latitudes in the coldest season of the year to watch auroras when it is so cold that even people's eyelashes get frozen. However, these ladies know that once the aurora arrives, the cold melts from your mind as the aurora captures your heart. It is a special experience seeing the aurora in the freezing temperature of –30°C in Canada's north. Photo by Yoshi Otsuka.

▲ Fig. 1-12. By luck or fate, a major solar flare occurred when a Japanese television station was visiting Yellowknife to make an educational TV program for families. The flare, emitting massive coronal eruptions into the interplanetary space, generated a large geomagnetic storm as well as impressive red auroras. Members of the television crew had pitched a tent on the lake to wait for the appearance of gorgeous auroras. Photo with the kind permission of Tuhru Okada.

From Greenland:

Greenland, an autonomous country under Denmark, is also the largest island in the world. It is six times larger than Japan, and 90% of its surface is covered with snow and ice, providing a dynamic contrast of colorful auroras against the whiteness of ice glaciers and rugged coastlines. Because of the country's high geographic latitudes and long nights, one can enjoy long hours of aurora watching.

Greenland is a special place to study auroras. Because of its very high latitude, dayside auroras can be seen here.

Figure 1-13 shows an American-made scientific radar, which serves as a tool for joint research by scientists from all over the world.

▲ Fig. 1-13. Aurora radar located at Søndre Strømfjord, Greenland. It is possible to estimate ionospheric current vectors from measured data of the electron density in the ionosphere. At this location, chemical contents in the upper atmosphere, such as stratospheric ozone, are measured. Photo by the Stanford Research Institute International.

From Iceland:

Iceland is an island located in the volcano zone which extends straight north from the center of the Atlantic Ocean. There, one can enjoy the numerous scenic lakes and hot springs amidst rugged terrain and beautiful auroral displays at night.

It is interesting to note that if we were to trace the lines of the Earth's magnetic forces from Iceland to the southern hemisphere, we will eventually reach the Japanese Antarctic station (i.e., Syowa Station). This means that people in Iceland and at the Syowa Station can enjoy the same shape, color and motion of auroras simultaneously.

Figures 1-14 and 1-15 show unique pictures of active auroras along the backdrop of some active volcanos in Iceland.

▲ Figs. 1-14, 1-15. Very active auroras dancing above Vatnajökull Glacier Lake, Iceland. These photos were taken just after the full moon rose, causing the snow to appear as white as if it was daytime. Photos by Tsukasa Enomoto. ▼

From Scandinavia:

In Scandinavia, the land of myths and folklore, one can enjoy auroras coupled with fjords, glaciers and hot springs. The Scandinavian countries of Norway, Sweden and Finland as well as Svalbard Islands are described below.

From Svalbard:

The Svalbard archipelago ("Cold Coast" in Old Norse) is farther north than most of Scandinavia. It is at least 2,000 km farther north than Oslo and 1,000 km farther than Tromsø, Norway. There are scheduled flights from both cities to Spitsbergen, one of the islands in the archipelago. The island's population is 3,000 people and wildlife are abundant.

Figures 1-16 through 1-19 show some of the unique auroras, mostly on the day side, taken in Svalbard.

▲ Fig. 1-16. Dayside auroras over Kings Bay, seen from the Chinese Yellow River Station (78°55'12"N, 11°55'48"E) at Ny-Ålesund, Svalbard. Photo by Ze-Jun Hu.

◀ ▲Figs. 1-17, 1-18.
These photos of rayed arcs were taken from the Kjell Henriksen Observatory (KHO) in Svalbard. KHO is the optical station for auroras. The vehicle seen in the photos is a belt wagon, which is the only way to get to the station in the winter. To the right of Fig. 1-17, one can see a part of the station itself. The plastic dooms are where the instruments are housed. Photos by Njål Gulbrandsen.

▲ Fig. 1-19. Photo of the EISCAT Svalbard Radar system. This photo was taken when a small substorm was in progress. We were running the radars at the same time, and it often caused some interference with the photo camera that would show up as bands of noise in the photos. Photo by Njål Gulbrandsen.

From Norway:

Tromsø, Norway, is the largest city in northern Scandinavia and is home to the University of Tromsø, which was the northernmost university in the world. It is also famous as being the birthplace of auroral research—one unique experience to sample is to observe auroras from a coastal tour boat.

Figures 1-20 to 1-23 are a collection of aurora pictures taken in or near Tromsø.

◀ Fig. 1-20. Panoramic view of the city of Tromsø. Tromsø is one of the most beautiful cities located within the Arctic Circle. The University of Tromsø, which is the northernmost university and is historically famous for auroral research, is located here. Source: Shutterstock / Lunghammer.

▲ Fig. 1-21. This photo shows the aurora and some sea fog covering the bridge between the island of Tromsø and the next island. The aurora shown is a weak diffuse aurora. Photo by Njål Gulbrandsen.

Fig. 1-22. Norwegian forests and auroras. Source: Shutterstock / Jamen Percy.

Fig. 1-23. Auroras colored with green and purple appeared over Ersfjord, Norway. near the Tromsø harbor. Source: Shutterstock / V. Belov.

From Sweden:

Kiruna, in northern Sweden, is a city of iron and steel industries. It is also the location of the Swedish Institute of Space Physics. There, auroras are observed both by means of ground instruments and satellites in space. This Institute has attracted many young researchers from around the world. Nearby Kiruna, the northernmost railway between Sweden and Norway was constructed in the early 20th century to ship iron from this "non-freezing" habor. This harbor never gets iced because of the mild Mexican Gulf stream, which brings relatively warm temperatures all year round.

▲ Fig. 1-24. A large-scale auroral arc over an incoherent scatter radar in Kiruna, Sweden. This is one of the radars the EISCAT Community constructed as a major international project for unveiling aurora processes. Photo courtesy of the EISCAT Association.

▲ Fig. 1-25. A very bright diffuse aurora that occurred near the horizon, taken from the Esrange Rocket Facility in Kiruna, Sweden. Photo reproduced with the kind permission of Marcus Lindh and Ella Carlsson Sjöberg.

▼ Fig. 1-26. Very beautiful and powerful auroras dancing over the Esrange Rocket Facility. This colorful aurora was captured near the launching tower and the wind measurement tower. Photo reproduced with the kind permission of Marcus Lindh and Ella Carlsson Sjöberg.

From Finland:

In this beautiful country, more than 70% of its land is covered in forests, with as many as 180,000 lakes supporting the natural ecosystem. Northern Finland near Rovaniemi is supposedly the home of Santa Claus. This city, amongst others, is well known for aurora viewing and winter activities including skiing, snowmobiling and reindeer sledding.

Figures 1-27 to 1-29 show a few small-scale but very beautiful auroral displays.

▲ Fig. 1-27. This series of active auroras was taken by Jiro Yokoyama, a tourist from Japan just after he had arrived at and checked in at a hotel in Yellowknife. He happened to find this gorgeous light show that night even before he had stepped into his room. After setting up his camera, he immediately rushed back to the hotel's parking area. Photo by Jiro Yokoyama.

◄Fig. 1-28. A legend from Northern Finland tells us that auroras are generated in the sky by "powdered snow scattered from a fox's tail". Photo by the kind arrangement of Esa Turunen.

▲ Fig. 1-29. This is more than a record of auroras. "A break up show and a Finnish forest" make a perfect collaboration, upgrading this piece from just a photo of an aurora into a work of art. Photo by the kind arrangement of Esa Turunen.

From Russia:

In Russia, which spans from off Scandinavia to the western border to Alaska, Murmansk is probably the center of auroral activity. However, only the northwestern part of the Russian border near the Barents Sea is under the auroral belt. The remainder of the belt crosses over arctic waters until its Alaskan border, completing the circle.

Figure 1-30 shows a large-scale dynamic aurora which visited the Murmansk harbor. Figures 1-31 through 1-35 represent an important record of auroras as well as scientific data from what is called an "Aurora Hunting Trip". The group which took these photos consisted of scientists, a journal editor, and technicians, and was headed by Sergey Chernous, a professor of the Polar Geophysical Institute in Russia. He has more than 50 years of experience in auroral research. This group visited many different places near the Kola Peninsula. A common interest of this group is to take high-quality photographs of auroras. They move in any direction they want, depending on solar activity and real-time conditions of the solar wind.

Fig. 1-30. A large-scale, dynamic aurora, mirrored in the beautiful harbor of Murmansk. Photo by Aleksandra Bulgan.

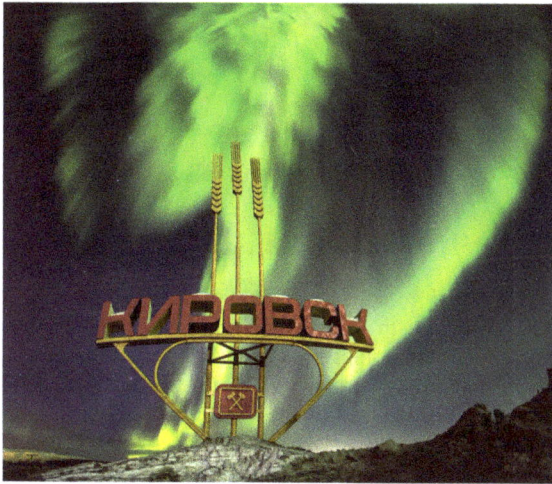

Fig. 1-31 to Fig. 1-35. Professor Sergey Chernous, space geophysicist from the Polar Geophysical Institute, with 50 years of experience in aurora research, is a world-renowned author on short-term forecasts of auroral occurrences.

Fig. 1-31. An aurora occurring over the entrance to the northern city of Kirovsk. As you can see in the photo, the aurora was very active. We had also successfully avoided capturing any trace of urban lights from Kirovsk.

Fig. 1-32. We had to wait more than two months to capture an image of the entrance to the second northern city, Apatity, under an aurora. Luck finally shone on the group, and we managed to photograph an excellent pink aurora with a greenish lower edge and red rays.

▲ Fig. 1-33. The word "Plateau", when translated from the language of the Sami, native inhabitants of the northern lands, refers to a "mountain with a flat top and long valley". That is where the group decided to take photos of active auroras, hoping to obtain good pictures of auroras with the distant lights of a mining settlement in the background. The picture we finally captured shows typical auroral displays of the auroral oval and rayed arcs, with a violet tinge. We call this kind of aurora a "sunlit" aurora because they get excited by solar light at higher altitudes (about 400 km above ground).

◀ Fig. 1-34. Glow over the water on a winter day, at Lake Imandra where fishermen were engaged in ice fishing in the extreme cold.

▲ Fig. 1-35. A panorama view of auroras over Lake Imandra. This lake is more than one hundred kilometers long. We saw many different forms of aurora that night.

1.2. Viewing Auroras from Antarctica

In the southern hemisphere, there is an auroral belt that is symmetrical with the auroral belt in the northern hemisphere.

From Antarctica:

The probability of auroral occurrence in Antarctica is the same as that in the northern hemisphere, but records are few and far between because of its low population density. However, there are many aurora observation stations in Antarctica. For instance, Syowa Station is located right under the auroral belt.

Figures 1-36 to 1-42 show the various types of localized auroras observed from near research observatories.

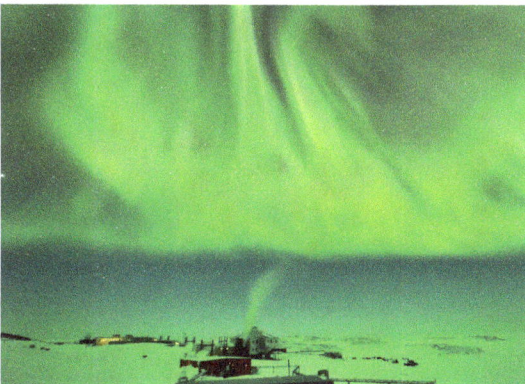

◀Fig. 1-36. A green aurora covering the entire area of the Japanese Syowa Station. Photo by Yasuo Takeda, the Japanese Antarctic Research Expedition.

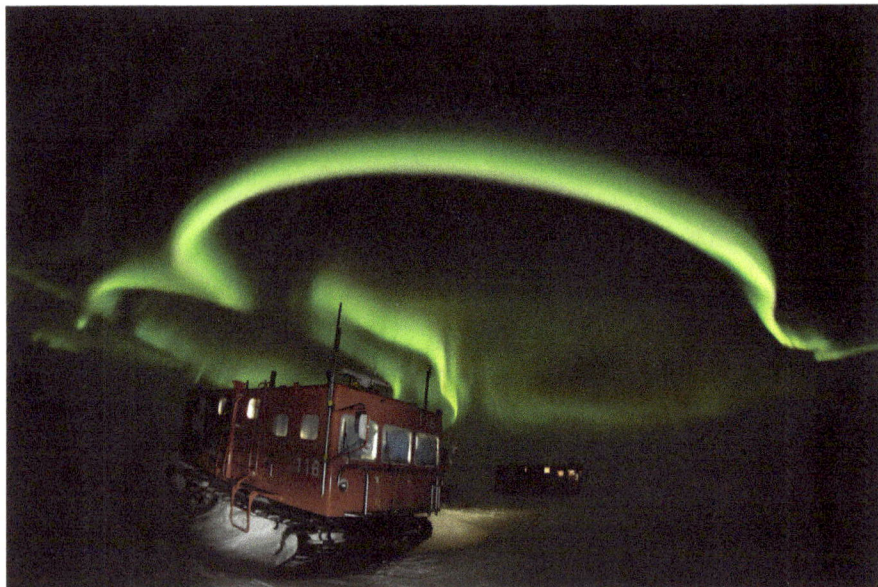

▲ Fig. 1-37. Auroral arcs, the simplest form of auroras, over the Syowa Station in Antarctica. Photo by the Japanese Antarctic Research Expedition.

▼ Fig. 1-38. An example of an aurora with a multiple arc system appearing above the Syowa Station in Antarctica. Photo by the Japanese Antarctic Research Expedition.

▲ Fig. 1-39. An intense ray structure, which represents the direction of the geomagnetic field, that appeared over the Syowa Station in Antarctica. Photo by the Japanese Antarctic Research Expedition.

▼ Fig. 1-40. Auroral spirals above the Syowa Station in Antarctica. Photo by the Japanese Antarctic Research Expedition.

▲ Fig. 1-41. An intense discrete aurora dancing above the Aurora Observatory (the green building) at the Chinese Zhongshan Station (69°22'24"S, 76°22'40"E) in Antarctica. Photo by Ze-Jun Hu, Polar Research Institute of China.

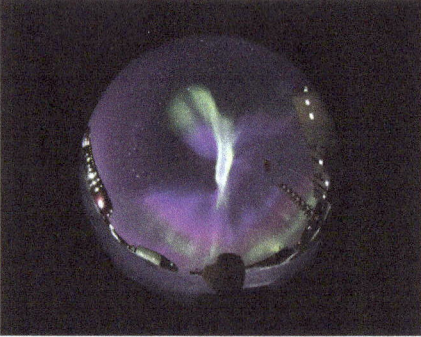

◄▲Fig. 1-42. Intense auroras over the Chinese Zhongshan Station (69°22'S, 76°22'E) at Larsemann Hills, Antarctica. Photo by Yang Liu, Polar Research Institute of China.

1.3. Viewing Auroras from Middle and Low Latitudes

Auroras are sometimes visible from the Earth's middle and lower latitudes.

From middle and low latitudes:

Some European cities, such as the UK and Germany, parts of the United States, Australia, New Zealand, and Japan, occasionally experience aurora activity during large geomagnetic storms—one may be extremely lucky to see an aurora while on holiday in one of these places: see Figures 1-43 to 1-48. These rare events occur only when the auroral belt expands considerably due to large geomagnetic storms and very high solar activity.

▲ Fig. 1-43. An aurora seen from Scotland, UK. Source: Shutterstock / Jim Hunter.

▲ Fig. 1-44. An aurora making an appearance over Boston, USA. Comet Hale–Bopp with a long tail is also seen in this photo. Photo by Frank F. Sienkiewics.

▼ Fig. 1-45. An aurora sighted from Bavaria, Germany. Source: Shutterstock / Jens Mayer.

▲ Fig. 1-46. Photo of an aurora over Rikubetsu Space and Earth Science Museum in Rikubetsu, Japan. Photo by Nozomu Nishitani.

▲ Fig. 1-47. Aurora sighted from Tasmania, Australia. Source: Shutterstock / Stephanie Buechel.

▲ Fig. 1-48. An aurora seen from Queensland, New Zealand on March 7, 2003. Photo by Toshio Ushiyama.

The color of such auroras at these latitudes is somewhat different from those at high latitudes, which are colorful and dynamic: see Chapter 4 for a deeper explanation

Auroras at mid- and low-latitudes are rare and typically red in color. Such auroral occurrences once convinced people that something unlucky was about to happen. This may be because red is the color of blood; some believed it symbolized that a relative of the person who saw a red aurora would soon die. Even till today, fire stations are often alerted whenever red auroras appear.

Chapter 2
Various Types of Auroras

Historically, auroras were classified according to their color, shape, motion, and so on. These classifications were often subjective, as they depended on factors like what the researcher logging the observations perceived, the tools used to observe the aurora, where and/or the angle he or she is viewing the aurora from, as well as prevailing weather conditions at the time of observation. Auroras were also classified into three main types (according to their morphology)—diffuse, discrete and pulsating auroras.

1. Diffuse aurora: As is clear from its name, the entire or a large part of the aurora is diffusive and thus difficult to definitively identify as any defined shape. It is often impossible to distinguish the diffuse aurora from the clouds that may be present at the same time as both show up as whitish green or gray plumes.

2. Discrete aurora: In contrast to the diffuse aurora, discrete auroras have clear forms such as curtains, coronas, and surge- and torch-like structures.

3. Pulsating aurora: Where a part of the sky or the entire sky lights up in a repeating "on/off" pattern.

2.1. Curtain, Corona and Surge

In this chapter, various forms of discrete auroras will be shown in the photographs, with a brief account about how their special shapes—arcs, bands and coronas—are formed.

The conditions required for generating the different forms of auroras are still not clearly understood. They can also be seen differently using different observation tools.

a. Curtain auroras

◀ Fig. 2.1. A curtain aurora. This photo was taken during a mild winter in 2016. The ice under the Yellowknife River Bridge had already begun to melt in early April. This seasonal change is a very dangerous time. The ice on the river can melt faster than the lakes' because the water is flowing. The aurora curtain approached from the other side of the bridge. The aurora waved across the bridge many times that night.

Curtain auroras are a very popular form of auroras. These auroras are generated by energetic electrons accelerating along magnetic field lines into the upper atmosphere, which result in structures that are similar to curtains. This aurora form can have single or multiple folds.

b. Coronal auroras

◀ Fig. 2-2. A coronal aurora photographed in late August in Yellowknife around 11 p.m.

The corona form results from auroral forms expanding from a point in a radial fashion. This shape is best observed when viewing vertical auroral curtains from directly below them. The point from which the lights are directed radially is called the coronal point. The central portions of a corona aurora often appear to reflect red or reddish-brown hues. This is because the upper part of each auroral 'curtain' is characterized by red auroras.

Looking at a large-scale corona aurora sometimes causes the watcher to feel as though one is being pressed downward into the ground. And because corona auroras occur at the zenith of an aurora formation, it is often followed by a sudden and dramatic breakup in the aurora, which resembles Aristotle's description of "the heavens being split apart".

c. Traveling surge

◀ Fig. 2-3. A traveling surge appearing over the Syowa station in Antarctica. Photo by Junji Horiuchi, from the 2003 Japanese Antarctic Research Expedition.

A "traveling surge" refers to large-scale dynamic motion of auroras traveling westward from midnight. This motion resembles a surge and its trail can stretch for as far as 1,000 km, at a speed of 1–2 km/sec—in other words, it can travel from the middle of Canada to Alaska in only 10 minutes!

2.2. Dynamic Motions/Various Forms and Color

The characteristics of auroras are not limited to their (rapidly changing) forms and colors; their dynamic motions, which progress without sound, are another aspect. Some examples of adjectives for the motions of auroras are "curl" and "drift".

Auroras move in sequences made up of a combination of these motions, producing a kind of complicated dance of lights. Aurora watchers have often described feeling a sense of wonder as they watched these ever changing— almost enigmatic—"light shows". Auroras have likewise caused scientists to wonder how much human beings actually understand the natural phenomena around us.

d. Various colors

Although auroras are commonly whitish green, they come in a variety of colors, depending on the energy of the particles precipitating from space and the altitudes from which they were generated. Interaction time, the time necessary to emit lights after precipitating particles collide with atoms and molecules in the upper atmosphere, is an additional critical factor that determines an aurora's general color.

Occurrences of red auroras are rare, and this rarity has resulted in folklore portending red skies as ominous omens foretelling an "unlucky" event [Fig. 2-4] Even now, whenever this phenomenon occurs, people residing in the polar region try their best not to look upward.

▲ Fig. 2-4. On March 17, 2013, a huge aurora event occurred after 3 a.m. It covered the entire sky in red, which can be seen with the naked eye. The color was so strong that many who saw it misinterpreted the event as city lights reflecting off a large cloud. Aurora events like this one have occurred throughout history, but not often. For many of the people who saw the red aurora in the photograph, they were experiencing it for the first time in their lives. This red aurora continued into the morning daylight.

2.3 Auroras and me: an essay by Yoshi Otsuka

Auroras in all forms are both a scientific and deeply personal experience. Such a description would not be complete without a personal perspective. The following section is a written narrative by Yoshi Otsuka, the photographer behind most of this chapter's photos, and the time he has spent watching auroras. After reading his stirring account, perhaps you, the reader, will want to experience the "Aurora Life" too.

Watching Auroras
(by Yoshi Otsuka)

I have been living and working in Yellowknife as an aurora guide for a long time. Remarkably, I still get goosebumps whenever I encounter an aurora explosion, no matter how many times I have seen one before.

The best moment of an explosion is very short and can end in just a few minutes. You will never encounter the same thing twice. When I think about that, I find myself wanting to take photos to preserve that moment, no matter what. So, even though I know it is better to look up rather than spend the time readying my camera, I constantly find myself thinking, "If I can just capture even a little of the aurora's explosive energy in a photo..."

It was in January 1997 when I first encountered an aurora. Prior to that, I had fortuitously chosen to travel to Canada to experience a life outside Japan. At that time, I did not even know there were such things as auroras!

While job hunting during Canada's winter season, I heard that Yellowknife was famous for dogsled races. As I learned more about the place, I found that it was also a famous town for auroras, although I had never seen a picture of an aurora. I could not imagine what an aurora would look like; only the words "South Pole" and "Light Curtain" came to mind in relation to the word "aurora". Since there was not much information available about Canada's North, I headed to Yellowknife via a long-distance bus. In my mind, I expected to see a small town with dogsled races, an extremely cold land, and indigenous people living there.

I saw my first aurora behind the clouds the night I arrived in Yellowknife. A strong light materialized in the clouds like green lightning, flashing. Soon after, I started working under a dog musher (a person who lives and looks after sled dogs during the winter); my first work-day was a day of many first experiences.

The kennel I worked for was located near a lake that turned out to be the best spot for viewing auroras. After work one day, I lay down alone near the lake to watch the aurora over the ice. With the sounds of dogs barking all around me, I witnessed a sudden aurora explosion, which encompassed the whole sky. I was overwhelmed by the furious light that blanketed the sky. I even forgot about the intense cold that had dropped below –30°C! It was an unforgettable day.

Since then, I have been attracted to the natural beauty of auroras and the extreme North. My interest has also expanded to include the culture and life of the First Nations, Métis and Inuit peoples living in the region. In spring that year, while traveling back to Yellowknife from a dogsled race, I witnessed the collaboration of an aurora and Comet Hale–Bopp. That celestial dance left a strong and lasting impression on me.

After my travels in Canada and returning to life in Japan, my experiences living in Yellowknife with its auroras never left my head. As a result, I soon returned to Yellowknife as a tour guide.

While working as a guide for a tour company, I began to think about how I could see as many auroras as possible without being restricted to a specific time or place. Hence, I started a small tour company for my own pleasure. I also hoped to share my knowledge on auroras and how to enjoy them with as many people as possible.

On almost all nights throughout the year, I am out on tours, so most of my photos capture the sights I saw with my guests. Whenever I look at these pictures, I reminisce—I remember the wonders that my guests and I experienced while watching auroras, as well as the beauty of the place from which we watched them. My pictures of auroras are my important memories and a record of having shared the experience with many people. The more I look at auroras, the more I know about them, and the more interesting they become.

When I started working on aurora tours, I tried to understand more about auroras by reading many books on the subject. It was during this time that I found many books by Prof. Yohsuke Kamide, which are the source of a lot of my knowledge on auroras. His writings were so interesting that I could not stop reading his books! They described everything from the basics to the details of the auroras I saw every night.

For example, his books taught me that an aurora is a result of plasma coming out from the Sun reaching the Earth's magnetosphere, resulting in an aurora reflected onto the sky in a tremendously magnificent scale.

Finding an aurora requires understanding as well as luck. An aurora changes its shape freely while moving in every direction. I observe the movement of each night's aurora to predict the time of its breakup. I also observe the movement of clouds while thinking about the time of the breakup, because they tend to obstruct one's view of an aurora. Cloud formation and movement is complicated. Sometimes, it is more difficult to predict than an aurora! If there are clouds between the ground and where in the sky the aurora occurs, one cannot see it even if it appears. I often ponder over the viewing conditions and think deeply, as if to put a puzzle together in my head.

That special moment when I am able to capture an aurora explosion with no clouds overhead is a most dramatic scene. There have been a number of times this happened while I was leading a tour group and I couldn't speak a word even though I am the tour guide!

Observers seeing an aurora explosion for the first time are often surprised by the speed of movement, strength of light, and intensity of color—all often manifesting in forms beyond their imagination. Sometimes, the sky may become so bright suddenly that it lights up the faces of the people around us even on nights when there is no moonlight. On nights with a full moon, an aurora explosion can commence without warning.

There is a wide variety of auroras to see and enjoy. For instance, there is one type whose light appears in bright green and pink and that moves actively, another that seems to pour downward like a shower, and one that resembles a red wall. Whenever people encounter such wonderful auroras, some scream while some others are at a loss for words.

I wonder how people in the past felt when they saw the northern lights. Had they been terrified by the explosions? Many local aboriginal people have told me interesting stories about the aurora, and how their people have lived in awe and respect of auroras since ancient times. Today, we can scientifically understand auroras and even see them in photos captured from the space, but even then, I cannot help but feel a sense of awe and respect for auroras, just like how the people in the past did.

I look forward to the appearance of an aurora every night before the sky gets dark. It will start to come out, somewhere on Earth, before evening begins, or sometimes as soon as the morning! An aurora can appear in Antarctica even when it cannot be seen in the northern hemisphere due to the midnight sun, so the auroras that I miss will be observed by someone somewhere else. And since most of us will not have the opportunity to see auroras from space, we have no choice but to wait for our next chance from one of the limited locations on Earth.

I am very lucky because I live in a city where I can see auroras frequently. Yellowknife is blessed with sunny weather and its latitude is relatively low among the areas where you can see auroras often. This also provides an opportunity to watch an aurora for a longer period in a year. Even so, someday, I hope to see and take pictures of auroras from other places around the world. I am sure that witnessing an aurora from a different viewpoint would have a different impression on me.

My wife says that I am an "auroraholic", because I am always thinking about auroras, even outside of work! In Japan, people say an aurora is "a thing to watch once in a lifetime". But I can safely say that once you have seen one aurora explosion, one will desire to witness it over again and again. Thus, I will continue chasing auroras while sharing my wonderful experiences with as many people as I can, for as long as I can.

▲ Yellowknife's seasonal change from fall to winter occurs from mid-October to mid-November. There are very few sunny days or clear nights at this time. On those few clear nights, wonderful auroras can appear easily. Many people stay up late on the weekends to enjoy the beauty of the northern lights then. I enjoyed taking pictures from my backyard on this particular day. People in my residential area were audibly exclaiming at the beauty of the aurora (photographed).

▲ This particular aurora was touching for my friends and me. It looked as if the light descended all the way to the ground. This aurora occurred one April, which is when the temperature usually rises and many friends start to visit each other as a result. As the cold was milder than that of winter, we talked all night about family, friends and our lives while watching great auroras.

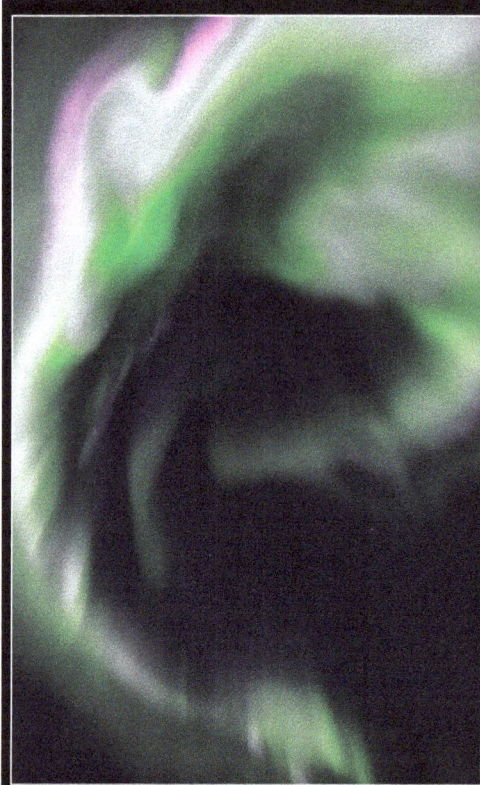

◀ Recognizing aurora colors with the naked eye is difficult for most people at first. It takes a little time for us to first get used to low-light conditions, and then see the aurora colors on top of those of stars. However, if the aurora is pink and very active, anyone can recognize it quickly. This type of aurora is very pleasing and exciting for tourists. People call out in excitement as pink aurora waves dance actively across the sky. ▼

▲ Two experienced world travelers who had seen many beautiful and amazing things around the world came to Yellowknife a few years ago. However, they were so impressed by the aurora they saw (see photos) that they continue to tell everyone they meet about their experience in Yellowknife. In their own words, it was "the most amazing experience they had had on their travels". As a result, many of these people are coming down to Yellowknife as well. ▼

▲ An amazing aurora had been expected on this night. I had prepared three cameras and was waiting on the frozen lake. The stars were beautiful and the aurora began close to the expected time. It was well beyond my expectation in its beauty. Auroras can surprise anyone.

Enjoying an aurora is a never-ending journey. Even on my days off from being an aurora guide, I still search for them if the weather is good. There are many things to learn about auroras. While we may have some expectations of their color or strength, shape and movement based on observable conditions, nature is complex, and auroras are not always predictable, which makes the experience more enjoyable. ▼

▲ The beauty of auroras reflected clearly on the water can be seen only for a short time during the end of summer and beginning of fall. As the weather gets colder, fog tends to rise from the lake, obscuring the reflection. Wind also tends to pick up with the season change, creating waves that distort the reflection.▼

▲ This aurora was very strong and bright—we are talking about almost glaringly bright colors. The aurora the day before had been lighter; the guests could only see white then, even though there were green, pink and purple hues in the aurora. This aurora appeared on the last night of their tour and they were so excited that they shouted: "The sky is burning!"

▲ During the coldest part of winter, we can drive vehicles on the frozen surfaces of lakes. A view from there gives us a clear open view of the aurora. In April, it is not safe to drive on the lakes due to warmer weather. Although we can still walk carefully on some lakes, we might forget to be careful when the aurora gets exciting.

◄ On 17 March 2015, two years to the day since the last red aurora, the night began with an aurora resembling the Milky Way. It was quiet for a little while, so we took a rest and returned to watch a very large red aurora from about 3 a.m. until dawn began to break.

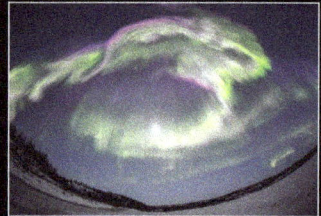

◄ On this evening, a wispy aurora was meandering through the stars as we were driving to the lake. I hurried to get there, and upon arrival, the light exploded from the sky. The light seemed to jump from the center and spread to all corners of the sky. Its energy is impossible to describe in words or images. I couldn't take my eyes off the aurora explosion and continued to press the shutter.

▲ A couple on their honeymoon requested for some aurora photos with their finest wedding apparel. We were able to have good weather and an aurora display that danced slowly at first and then faster and faster until it exploded. The couple also moved along a small island in the river as the aurora danced above. We got many pleasing photographs that day.

▲ We were having a forest barbeque while waiting for night and an aurora viewing. Suddenly the intense aurora appeared in the sky just after sunset. It was surprising to see it so well with the naked eye when only a handful of stars were visible. What a colorful dinner show we had.

▲ During a very active aurora season, many local people became more interested in auroras. They began driving near tour groups along the ice roads and lakes outside of town. On one of the evenings with a good aurora forecast, there were many vehicles with people waiting to see the aurora. As the night went on, some got tired of waiting and gave up for the night. Those that stayed and waited for nature to take its course saw an amazing aurora explosion.

▲ We were at the end of the road in Yellowknife and the beginning of the north. To go farther, we would have had to fly or make an ice road in winter. The ice road to the diamond mines is usually open in February and March. It is very far from Yellowknife, and there are no services, but there are often nice auroras there. Even commercial truck drivers must feel like stopping to enjoy their beauty.

▲ The strong solar wind was bringing a beautiful aurora at the same time the clouds were creating a patchwork quilt over most of the sky. I could find some clear spots here and there, but the timing was critical. In a miraculous moment, the clouds separated as a red wall-like aurora appeared before us.

▲ Some of us possess eyes like a camera. But on this night, our human eyes could not find the distinction between the faint aurora and thin clouds. Meanwhile, the aurora spread like a fog through the sky until our cameras could catch the purple curtain traveling above us.

◀ The sky in Yellowknife seems wide open and large because it has no mountains to block the view of the stars or other celestial beauty. While waiting for aurora, you can enjoy a 360-degree view of the stars, moon, Milky Way, satellites and shooting stars in the crisp clear air. Often you can see animals doing the same, such as owls and wolves. You can see so far that you can see the aurora swirling and approaching from the horizon.

▲ An aurora can change its shape in an instant. This bow tie shape appeared overhead suddenly and was captured by an opportune shutter click.

◀ ▲ This aurora explosion appeared like countless arrows of light raining from the sky. It is easy for one person to see this display while the person beside them sees nothing at all. This type of aurora, with such movement and color, feels like it is a living being sending a message. If you were cleaning your lens at just that instance, you would have missed it.

▲ In my family, the aurora is central to our lives in more ways than one. I took my daughter to see her first aurora when she was 3 months old. At 1 year old, she could recognize and point to the aurora in the sky. The beauty of a child and an aurora are both amazing.

◄ "I don't know where to look! I want three eyes! My neck hurts!" are a small sample of exclamations I have heard from guests over the years, whether or not it was their first time at an aurora viewing. Such sentiments are normal because a huge aurora explosion spreads all over the sky so extensively that it is impossible to see the whole picture at once or—needless to say—take a picture that captures everything.

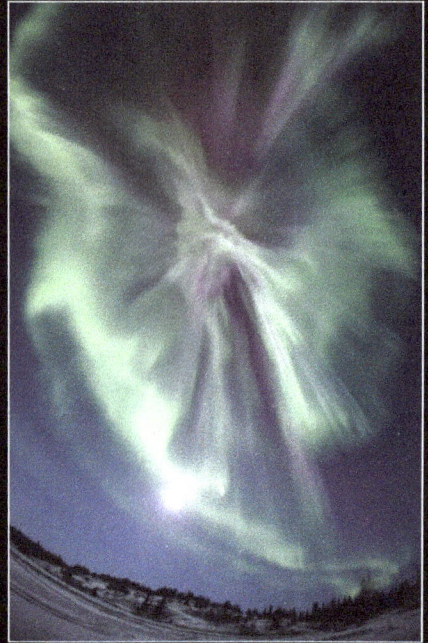

◄ The tall curtain slowly approached and finally reached us. The center of the explosion looked purple. The sight of an aurora spreading slowly and covering the sky sometimes gives me a feeling of fear.

▲ In autumn near Yellowknife, clouds may suddenly come close and disturb the appearance of an aurora. One night an aurora explosion happened at the boundary of the clouds and the dark blue sky. It was a small open space, but fate smiled on us as the explosion occurred on the clear patch rather than on the cloudy side.

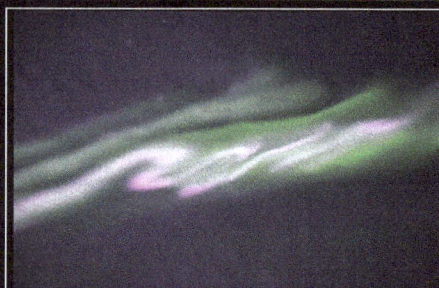

▲ I always wondered why an aurora display turns into a particular shape and why it moves in a certain way, for instance, in the form of "the curtain waves" or "light that flows like water", and the shape changes continuously. One will never get bored of seeing an aurora as every aurora and aurora experience is different.

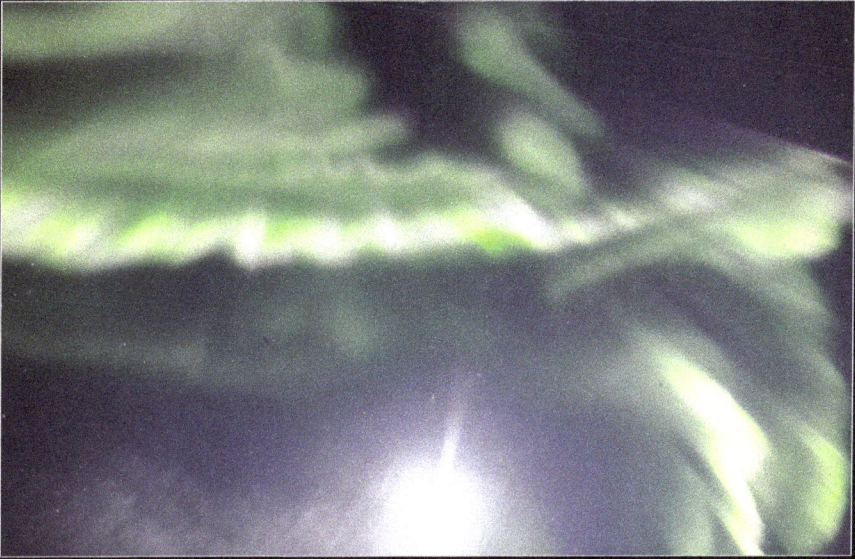

▲ Sometimes there are cases when I felt that the explosion happened out of the blue, such as on this occasion when I was watching a fascinating full moon, an aurora explosion suddenly began. I could not see the thin aurora behind the light of the full moon (though my camera successfully captured it).

On this occasion, the aurora came from the other side of the horizon, transforming itself into what looked like curtains with layers covering the moon. On days like this, I sometimes feel like I am wrapped in an aurora and will shortly be transported into outer space when it spreads across the sky. It's easy to forget that the aurora is actually 100 km above me! ▼

▲ When I was out walking in the forest one night, the light of the aurora suddenly got stronger and covered the sky. It appeared like a dragon flying across the sky, whirling, and then going away. Asians often compare auroras to dragons.

◄ On this day, I was shooting an aurora that was constantly moving and changing into many shapes. It finally came to rest over a group of trees, covering them like a circlet. It stayed in that form for some time. Was the aurora entertaining me on purpose?

▲ I wonder what this man was thinking about as he stood between the night sky and the aurora reflected on the lake. It must have been the moment when one of his dreams came true; his fascination with auroras has brought him back to Yellowknife every year since then. If one ever encounters an aurora explosion, it will be your most memorable event ever.

The aurora is something to experience rather than to watch. No matter how many times you see photos of auroras, it does not mean that you have experienced them. These are special memories that can only be shared with people who have gone through the same experience as you. ▼

▲ Beyond just hope and luck, we hunted for the small holes of deep blue in the clouded sky, catching one just in time as the aurora appeared. The aurora seemed to sweep the clouds from the sky, and our memories as well. We were all speechless, amazed and impressed by the sky full of stars and the aurora, without understanding how it happened.

People have rarely seen such an amazing aurora, even in a lifetime of living here. We traveled half the world to arrive at this spot at this lucky point in our lives, and we managed to experience such a rare and beautiful aurora. That is amazing.▼

Chapter 3

Auroras Seen From Space

If intelligent beings on other stars deep within the universe were to approach our solar system, they would first come across two rings of lights that surround the Earth's north and south poles. Though we can see things within visible wavelengths on the Earth's surface, these intelligent beings may be able to identify wavelengths of electromagnetic waves, which we cannot. Thus, they would be able to see the auroral belts by detecting radio emissions of several hundred kilohertz (kHz). These wavelengths are called auroral kilometric waves.

Emissions from the Earth would be equally detectable in the same way we identified radio waves coming from Jupiter. Through these electromagnetic emissions, intelligent beings would be able to discover that the Earth rotates in a 24-hour-period and that it is the third planet of our solar system.

Intelligent alien beings, through auroral waves, would also discover an auroral belt on Saturn, but fail to see one on Mars or Venus. Furthermore, they would soon find that although the Earth's auroral belts are primarily green in color, the auroras on Jupiter are reddish.

The Earth's two rings of lights are auroral belts encircling the north and south geomagnetic poles, which means astronauts can enjoy the spectacle of auroras from the International Space Station (ISS) that is located about 400 km above the surface of the Earth. Astronauts who have spent at least several days in space have described auroras as "wavy flames emitting music".

By using high-sensitivity cameras onboard space vehicles, one now can photograph auroras from space. It is now also possible to record the entire forms and dynamic motions of global auroras from above the auroral belts. Previously, it had not been possible to capture the large-scale auroral distribution due to the handicaps of ground-based cameras (which were for a time the only tools available to capture auroras).

In this chapter, we will show a number of auroras from various altitudes in space.

3.1. The International Space Station

The International Space Station (ISS), conducts a variety of experiments under gravity-free conditions. This altitude interestingly corresponds to the altitude where the highest point of the upper part of auroral curtains is found.

Figure 3-1 shows the appearance of the ISS. It was launched into orbit in November 1998, and it weighs more than 400 tons. It is no surprise that it is the largest man-made body in space, measuring 73 m in length, 108 m in width, and 20 m in height.

▲ Fig. 3-1. The ISS, located about 400 km above the surface of the Earth, is a joint project of five space agencies: NASA, Roscosmos, JAXA, ESA and CSA. Credit: NASA.

To date, the ISS is the only man-made vehicle that can fly inside auroras. This means that astronauts onboard the ISS have the enviable experience of looking into the most intense part of auroras. From the windows of the ISS, one can see faint auroras begin to show up from the horizon and approach it, all the way through to an explosive brightening known as the aurora break up—and the beauty of the latter scene is unforgettable.

Here is a selection of pictures of auroras as seen from the ISS, and a space shuttle.

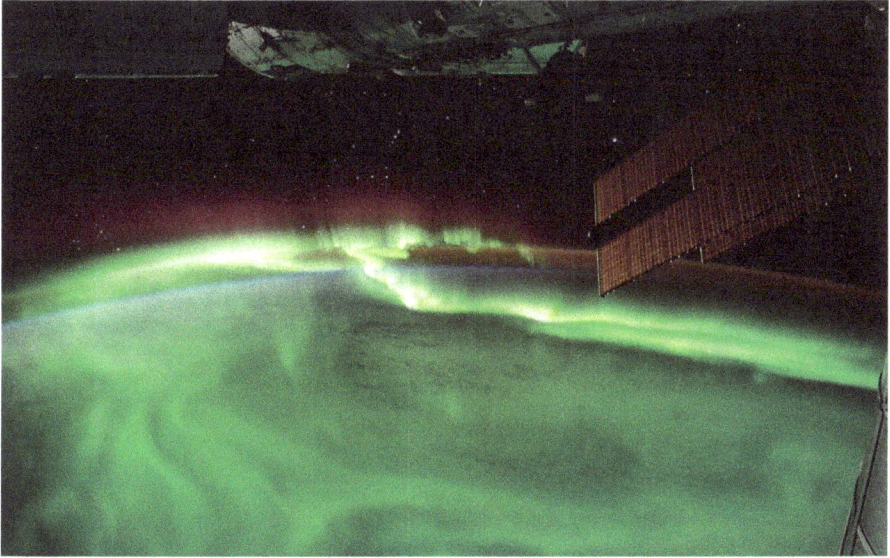

▲ Fig. 3-2a. The ISS traveling over the southern hemisphere as if it dives into the sea of auroras. Discrete auroras are seen in the middle and to the right, and pulsating auroras occupy the bottom half. The upper part of the discrete auroras is tinged with a reddish color. This picture was taken by the crew of Expedition 29 of the ISS on September 17, 2011. Credit: NASA.

Courtesy of Image Science & Analysis Laboratory, NASA Johnson Space Center

▲ Fig. 3-2b. Auroras photographed over a backdrop of city lights from the Midwestern United States, including Chicago (at the center), as well as St. Louis (lower-right), Minneapolis–St. Paul (middle-left), and the Omaha–Council Bluffs metropolitan area (lower-left). This picture was taken by the crew of Expedition 29 on September 29, 2011. Credit: NASA.

▲ Fig. 3-2c. The ISS skimming discrete and pulsating auroras. The upper part of the discrete auroras is tinged with a reddish color. This picture was taken by the crew of Expedition 29 at a point over the southeast Tasman Sea near southern New Zealand on September 17, 2011. Credit: NASA.

▼ Fig. 3-2d. Fantastic discrete and diffuse auroras captured with the moon and city lights from Helsinki, Finland. The exposure time is about 2 seconds, making the city lights elongated because the ISS moves at a speed of 27,600 km/hour. The exquisite contrast of the color changing from blue, to green and then purple from the ground up is so wonderful. The greenish auroras are supposed to move rapidly. The green color comes from atomic oxygen, whereas the blue and purple colors most likely come from nitrogen molecules. This picture was taken during a magnetic storm of August 31, 2005. Credit: NASA.

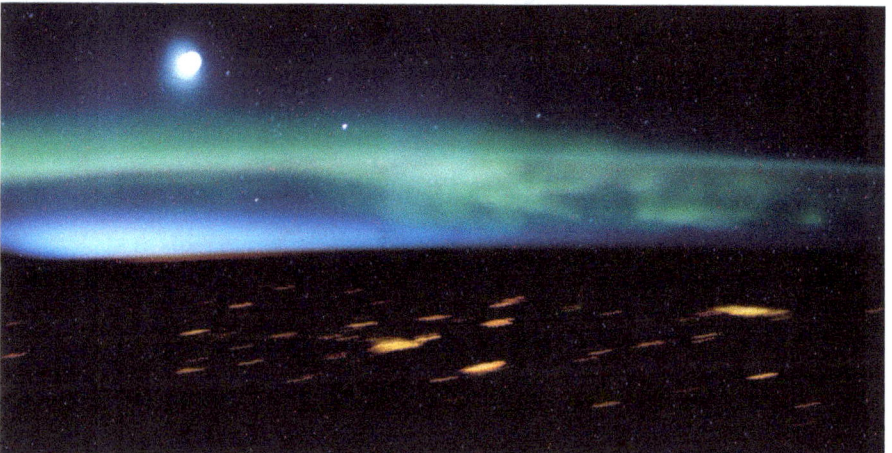

▲ Fig. 3-2e. Auroras colored green at the lower part and red at the higher part. This picture was taken by the crew of Expedition 31 in a magnetic storm on May 22, 2012. Credit: NASA.

▼ Fig. 3-2f. Spectacular auroral curtain, which looks like a wavy ribbon. Small-scale curtains are embedded in the curtain near the bottom-left corner. Rays are also found near the bottom-right corner. This picture was taken by the crew of Expedition 23 in a magnetic storm on May 29, 2010. Credit: NASA.

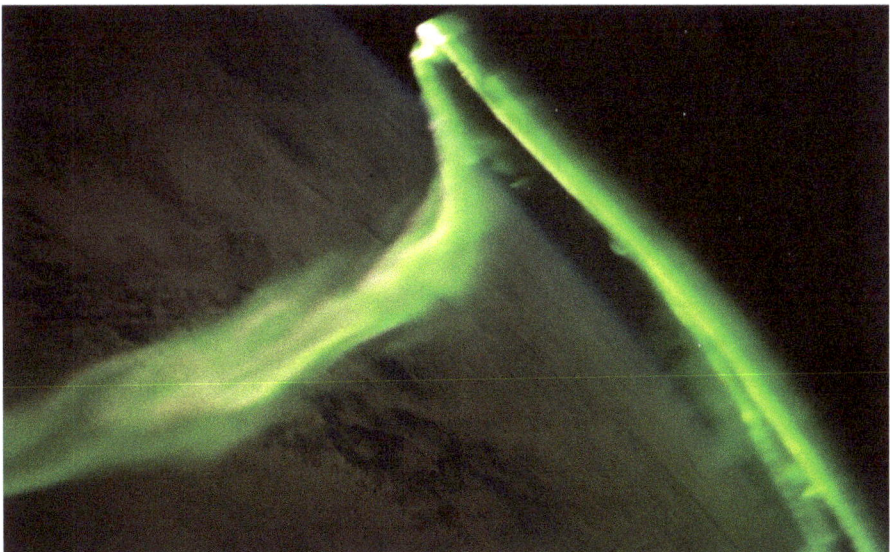

▲ Fig. 3-2g. The ISS going over what looks like an auroral "sea". The sight of many rays extending upward from the bright parts of the aurora is impressive. The Canadian robot arm is visible in the center. This picture was taken by the crew of Expedition 32 near the peak of a magnetic storm on July 15, 2012. Credit: NASA.

3.2. Beginning of a Space Era of Aurora Observations From Above

International Satellite for Ionospheric Studies 2

The year 1971 marked an epoch-making time in the history of auroral research. The International Satellite for Ionospheric Studies (ISIS) program is a joint mission between Canada and the USA. It was through the ISIS program that a group at the Physics Department of the University of Calgary, Canada, made successful observations of aurora distributions from ISIS-2, a polar-orbiting satellite 1,400 km above ground.

Although its timing accuracy was still primitive, ISIS-2 contributed enormously to auroral research, which had been relying on ground-based all-sky observations. Before this satellite, composite images created from ground-based all-sky camera data were the only means available to identify auroral distribution on a global scale.

It was also through ISIS-2 that the so-called diffuse aurora, which had been notoriously difficult to capture from all-sky cameras installed on the ground, was discovered.

There are two types of large-scale auroras: discrete and diffuse. As the terminology implies, a diffuse aurora is, in general, weaker than a discrete aurora, but it reflects the large-scale structure of the magnetosphere. How these two types are related is a common concern of auroral research. The group at the University of Calgary led by Dr. A.T.Y. Lui suggested that a diffuse aurora signifies the magnetosphere during magnetically quiet times. More strongly, by studying the spatial structure of the diffuse aurora, one is able to obtain the structure of the plasma sheet, which is the source of a diffuse aurora.

Figure 3-3 presents an example of a diffuse aurora captured by ISIS-2. The diffuse aurora is elongated in the east-west direction equatorward of vortex-like structures.

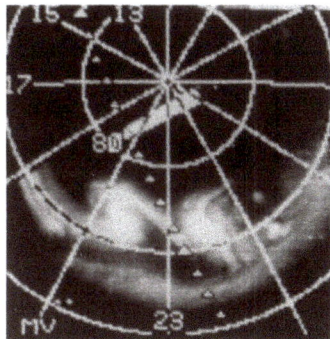

◀ Fig. 3-3. Example of a diffuse seen by ISIS-2, which is seen equatorward of the vortex-like structure of the aurora. The numerical figures "13", "15", and "17" represent the local time, and the figure "80" represents the magnetic latitude. The triangles indicate the positions of ISIS-2. Courtesy of C. D. Anger.

The Defense Meteorological Satellite Program

The Defense Meteorological Satellite Program (DMSP) is a meteorological satellite that measures the surface conditions of the atmosphere to facilitate weather predictions. Similar to ISIS-2, this series of polar-orbiting satellites provide us with global imagery of auroras per orbit, which takes about 90 minutes. However, it can also catch details of auroral structures, which change with substorm electron dynamics.

Presently, the DMSP series of satellites not only carry photometers for visible and ultraviolet wavelengths but also the spectrometers for capturing the intensity of precipitating auroral electrons. The excellence of the detectors from the DMSP series has seen its production line continue from DMSP-F1 up to DMSP-F19.

Figure 3-4 shows auroras and city lights of the United States taken by DMSP. Multiple arcs over the southern part of Greenland and Iceland are shown in Fig. 3-5.

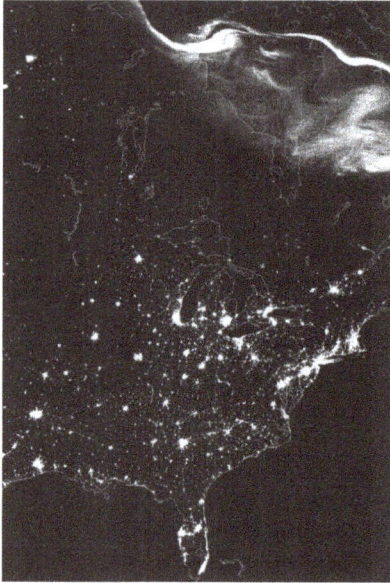

▲ Fig. 3-4. A black and white image of North America and an east-west elongated aurora seen by the DMSP. The aurora looks like a writhing snake. City lights in United States (for example, Chicago, Houston, Boston, New York, and Washington, D.C.) and Canada (for example, Toronto, and Ottawa) are clearly identified. Courtesy of H. W. Kroehl, NOAA/NGDC.

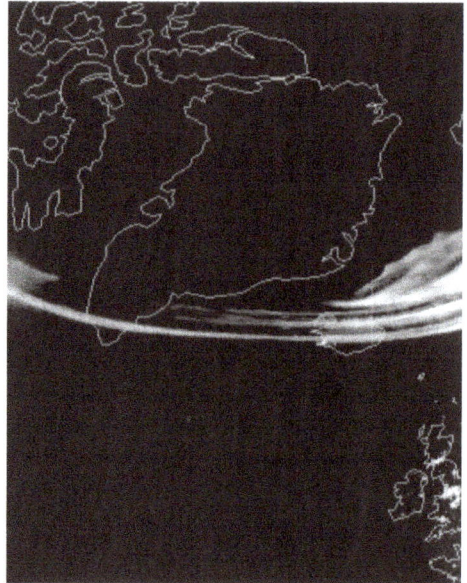

▲ Fig. 3-5. Multiple arcs over the southern part of Greenland and Iceland recorded by the DMSP. Courtesy of NOAA/NGDC.

3.3. Auroras Seen from Higher Altitudes

Dynamics Explorer

The ISIS-2 and satellites from the DMSP (up to time of writing) can only reconstruct one picture of the auroral distribution in one orbital period. This was because they can only conduct horizon-to-horizon scans, and an entire scan of the polar region takes approximately 10–15 minutes to complete, thus making it difficult to capture rapid changes in auroras.

Dynamics Explorer 1 (DE-1), launched by NASA in 1981, solved this shortcoming. From an altitude of 20,000 km, the DE-1 could take aurora photographs once every 12 minutes. Thus, we were able to observe the entire auroral oval expansion equatorward. An example of the whole picture of the auroral ring is shown in Fig. 3-6.

Observations were made that an auroral arc near midnight could suddenly change—for example, the enhancement of the arc would lead to a so-called

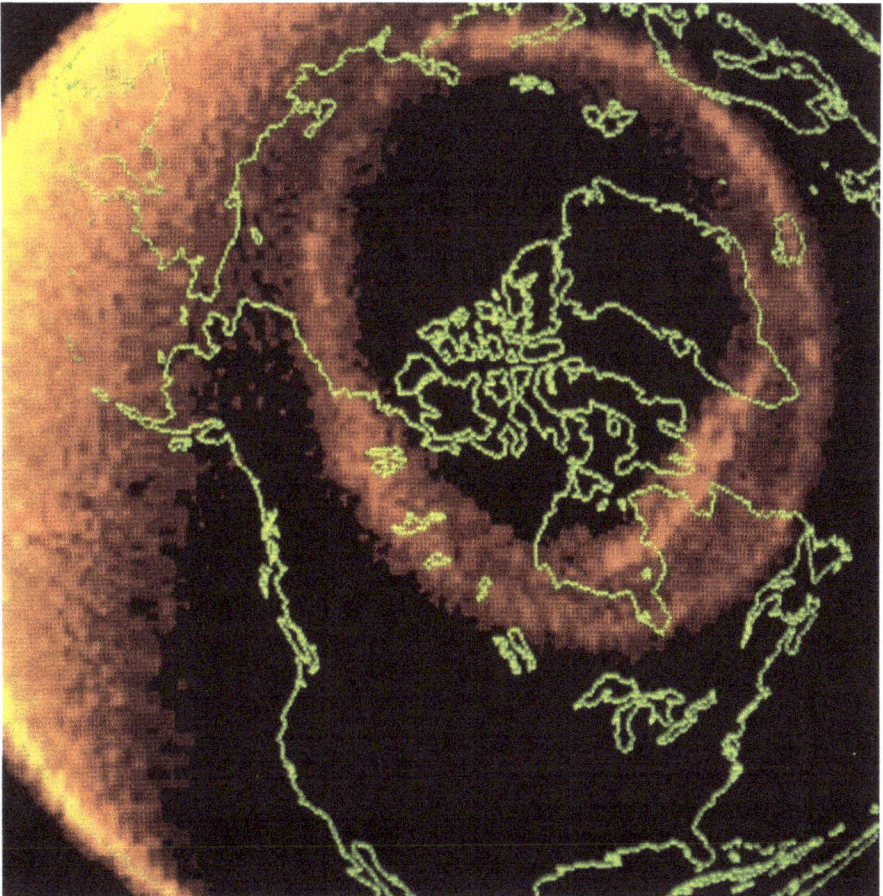

▲ Fig. 3-6. North America and whole distribution of the auroral ring captured by DE-1 at an altitude of 20,000 km. The left-hand side of the image is bright because of sunlight. Namely, the Sun is to the left. The green lines indicate coastlines (artificially drawn). Courtesy of L. A. Frank.

"expansion onset", which would signal the occurrence of what is called an auroral substorm. The observation covered the entire process.

When Prof. Louis Frank of the University of Iowa presented an early result from DE-1 at an American Geophysical Union's meeting, I (Kamide) thought that most of the existing issues regarding substorms would finally be solved, since the DE-1 could monitor the entire polar region for nearly three hours—the period during which substorm energy is stored and suddenly released.

One other auroral feature that the DE-1 discovered was the theta aurora, so named because it resembles the theta, a Greek letter (Fig. 3-7). It typically occurs when the interplanetary magnetic field is directed strongly northward. On such a rare occasion, the solar wind and the magnetosphere couple in a unique way, such that magnetic field lines in the polar cap are closed.

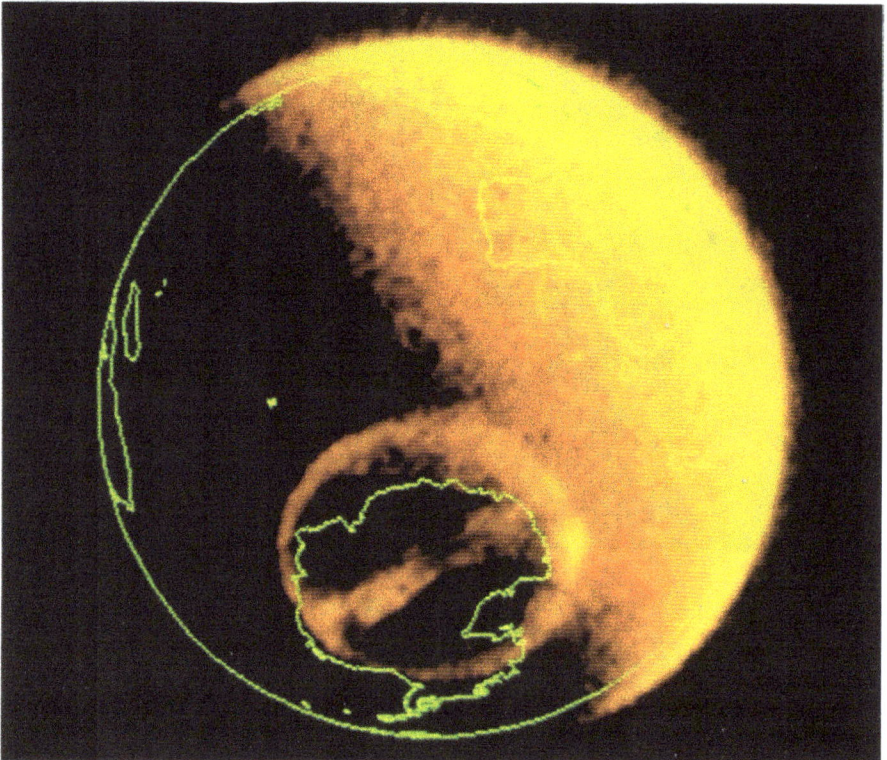

▲ Fig. 3-7. Auroral ring whose size is comparable to the Antarctic Continent. Usually, auroras are absent in the polar cap, which is the region encircled by the auroral ring. A new type of aurora was discovered, which connects the day side and the night side with a straight line. Because the shape resembles the Greek character θ (theta), this type of aurora is called a theta aurora. Courtesy of L. A. Frank.

Viking

This Swedish satellite, launched into orbit in 1986, was able to watch auroral activity from space with a 20-second time accuracy. Unfortunately, its lifetime was rather short—only a year. This satellite was also unable to capture the fine structures of auroras during many auroral substorms.

One of the Viking's strongest points was that it took many more photographs compared to ISIS-2 and DMSP, which can only take one photograph per orbit. It was thus possible for Viking to capture the heart of an auroral breakup, which led to the coining of the term the "eye of the aurora", just like "the eye of a cyclone, typhoon, or hurricane".

In the midnight meridian, a sudden brightening of a preexisting auroral arc and its rapid poleward motion signal the beginning of an impressive light show, which typically lasts for about 30 minutes or so. Scientists compare the variations of this light show with satellite particle data to account for the processes that could generate the characteristics of auroral dancing.

Figure 3-8 shows a typical aurora breakup seen from Viking. It is clear from the photograph that a breakup begins in an east-west elongated shape in the midnight sector. It then develops mostly westward toward the evening. If one or more breakups occur, subsequently, the brightening area would become more complicated.

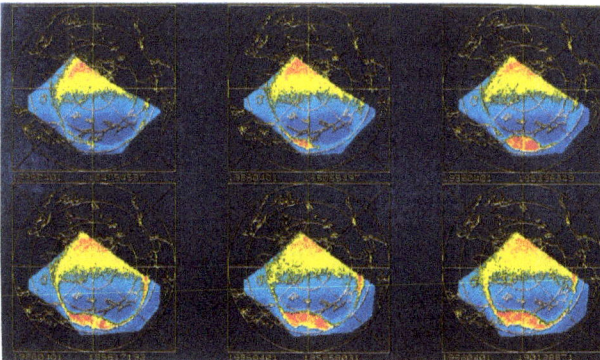

◀ Fig. 3-8. An auroral breakup seen by Viking. The Sun is to the top. The images were taken about once every three minutes. The yellow lines indicate coastlines (artificially drawn). The aurora became brighter in the east of the Scandinavian Peninsula, near the Arctic coast of Siberia. The bright aurora scontinued to spread wider and wider. Courtesy of R. D. Elphinstone.

Akebono

Japan's contribution to photographing global auroras from space was made by its Akebono satellite, which was launched into orbit at 15,000 km above ground in 1989. The highly sensitive UV camera on board was able to capture both global and local aurora features, particularly during the period leading to auroral brightening and its subsequent expansion.

An auroral bulge, another characteristic feature of a breakup, was nicely captured by the UV imager onboard the Akebono as shown in Fig. 3-9. This identified a stepwise development of auroras in the westward direction. It was found that the development of bright auroras associated with the stepwise growth of the westward electrojet occurred in a timescale of two minutes. This characteristic growth occurred in conjunction with the poleward and westward development of the substorm disturbed region.

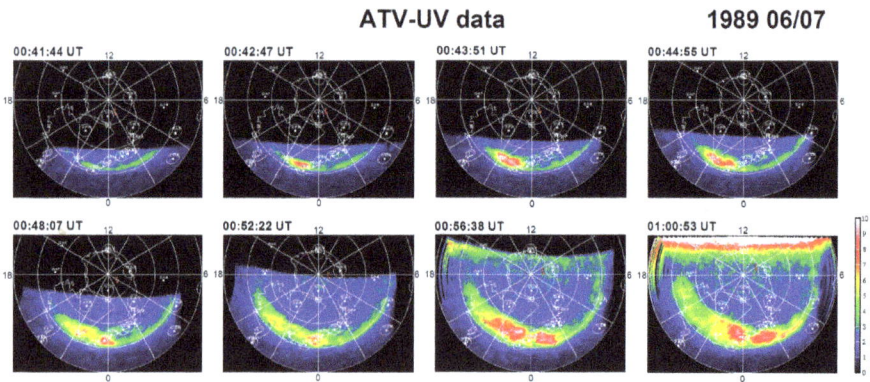

▲ Fig. 3-9. Auroral breakup seen by Akebono over Antarctica. The Sun is to the top. The white lines indicate coastlines (artificially drawn). The aurora became brighter near Syowa station, Antarctica. See how the bright aurora spreads out over time. Adapted from Kadokura et al., (2002).

Polar

It is understandable that because the apogee of satellites in the magnetosphere is far distant, they will be able to see the entire lifetime of a substorm from beginning to end. On the other hand, however, this distance results in the satellites being unable to see the finer details of small-scale structures, which are the characteristics of substorms. This shortcoming has finally been solved, in a sense, by the launch of the Polar satellite in 1996.

The Polar satellite can continuously monitor auroral activity in the entire polar region for more than 10 hours from a height of 57,000 km. This satellite carries two excellent cameras, one for the visible wavelength range and the other for the ultraviolet range. One might say that computer technology and camera techniques for aurora shooting have reached a peak here!

Figure 3-10 shows an image captured by Polar showing how rapidly the entire auroral ring responds to very large geomagnetic storms. This was taken during a historically unforgettable storm, during which the auroral oval extended to the latitude of Florida.

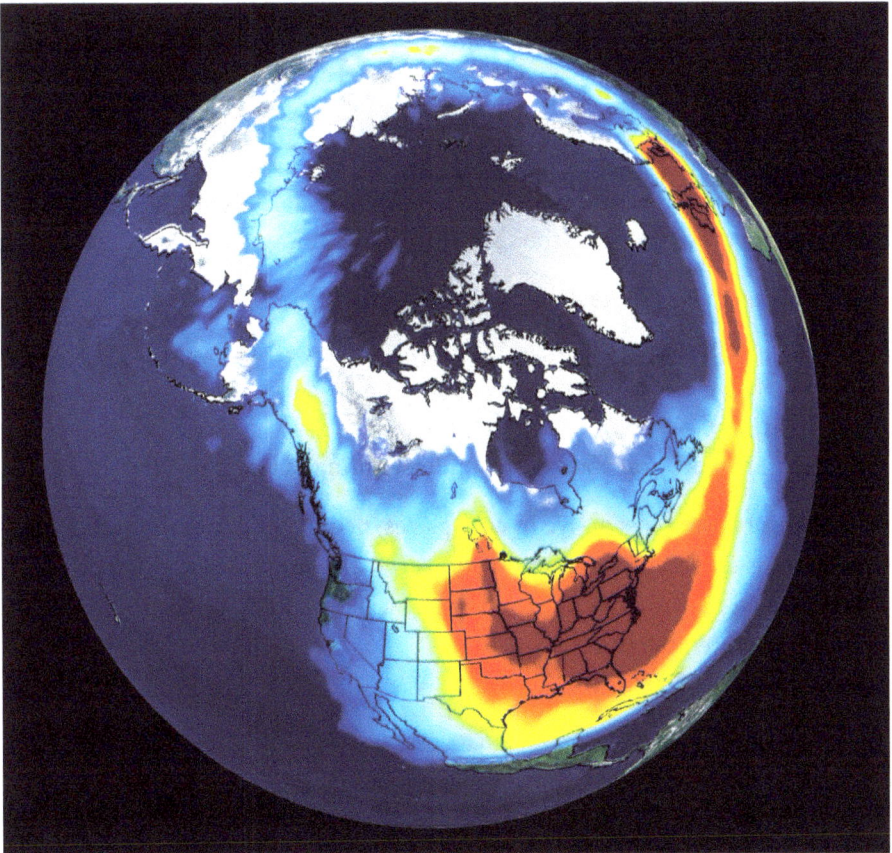

▲ Fig. 3-10. Extremely expanded aurora seen by Polar over North America in a large magnetic storm on July 16, 2000. The land and sea are artificially drawn in this image. Credit: G. Shirah, NASA.

Chapter 4

Unveiling the Secrets of Auroras: Science for Beginners

This chapter touches upon some of the basic questions regarding the occurrence of aurora lights. Here are some examples of such questions:

- How are the different colors of auroras created?
- Why do they move?
- On what physical principle do auroras simmer in the polar sky?
- From where do auroras originate?
- What part of space do the seeds of auroras come from?
- Through what route do these particles come, and how?
- What is the main role that these particles play in generating auroras?

There are, however, all sorts of misunderstandings as well.

Since auroras are most active in cold places, people tend to assume that a low temperature is a necessary condition. However, as you will learn in this chapter, the aurora is an electrodynamic phenomenon. Its production has nothing to do with the ground temperature.

4.1. What Are Auroras?

Auroras are visible dancing lights in the sky that are caused by the Sun's energy touching our atmosphere. In short, charged particles coming from the Sun through space collide with atoms and molecules that constitute the Earth's upper atmosphere. This collision results in the emission of light energy in the form we know as aurora lights. The basic mechanism can be compared to that of fluorescent lamps at home and neon signs in entertainment areas. The different colors of auroras depend primarily on the energy level and speed of incoming particles, as well as the atoms and molecules that are excited by the energy collision.

People often talk about auroras in terms of watching television at home. It is as though we are looking up at a giant monitor or screen in the sky on which television programs are repeatedly projected. In the case of auroras, the "television programs" are "broadcast" from space (instead of television stations). Flows of charged particles, called solar wind, leave the solar corona (the outermost part of the Sun's atmosphere) and hit the Earth's sky.

We enjoy various "programs" that are prepared by solar wind and the Earth's atmosphere. At home, television programs are made at television studios. In the case of natural "television", the aurora "shows" are created "live" by interactions between solar wind, the magnetosphere and upper atmosphere.

Scientists who study auroras not only enjoy the shows produced by nature but also attempt to understand the relationship between the Sun and the Earth. By finding out more about the various colors, shapes and motions of auroras, we begin to learn and ask more questions.

There are a variety of colors, forms and motions in auroras, and their forms can rotate both clockwise or counter-clockwise with no set direction. Not to mention, auroras can also suddenly disappear or reappear, seemingly for no obvious reason. Then there are also "quiet" types of auroras, which can remain still for hours, unlike the more active and quick-moving auroras mentioned in earlier chapters. It is mysterious yet impressive to know that nature provides us with many different types of auroral displays to watch from the Earth.

4.2. Where Can We See Auroras Most Frequently?

Many, if not most, people think that the higher the latitude, the more likely one will get to see an aurora. This is not true. For example, if one goes all the way up to the poles, the probability of seeing auroras drastically decreases. The most likely place to view auroras is the auroral belt (or zone).

How often can people see auroras from under the auroral belt in the northern hemisphere? Statistical data sets indicate that auroras occur in the northern auroral belt about 240 days per year, weather permitting (see Fig. 4-1). However, the auroral belt in the southern hemisphere has not been well studied, possibly because the population there is not as high as in the northern hemisphere (so there are less people to see and/or report aurora sightings).

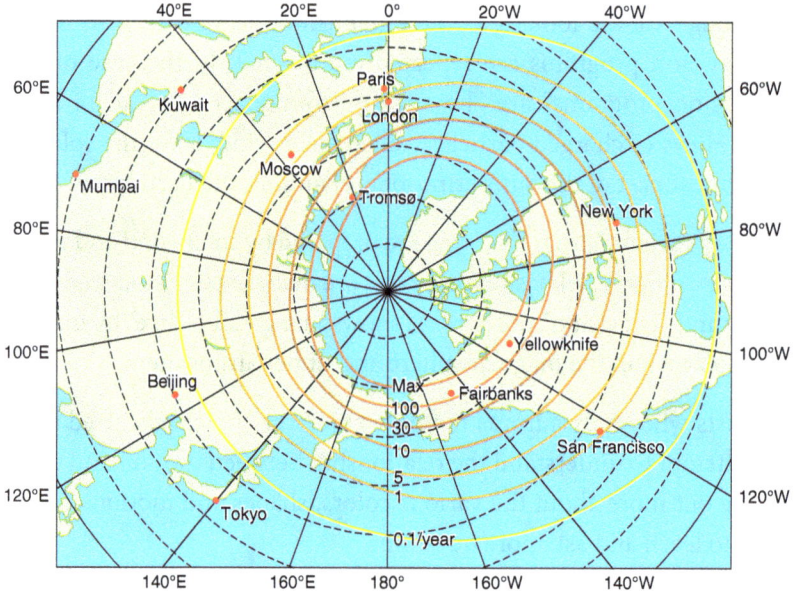

▲ Fig. 4-1. World map displaying one's probability of seeing auroras at different parts of the world. For example, the contour line labeled as "100" indicates where auroras can be seen about 100 days per year. Reconstructed from Hermann Fritz (1830–1883).

In any case, the Earth has two rings of lights surrounding the north and south geomagnetic poles. The Earth itself is a gigantic magnet. Everybody knows that the "N" and "S" of a compass point to the north and south poles, respectively. Nothing is strange about this. But scientists say that this is because the (magnetic) south pole is at the North Pole, and vice versa. This is confusing for people. How can the south pole be at the North Pole?

The geomagnetic pole can be defined as the point on the Earth's surface where the dipole magnet intercepts (meets) the surface of the Earth. The Earth is approximated by a dipole magnet. In mid-Europe and Asia, the difference between the geomagnetic north and geographic north is only 5 degrees or so. At higher latitudes, that difference becomes greater.

In Alaska, for example, the difference amounts to about 30 degrees. At high latitudes, there are places where the difference becomes as large as 180 degrees.

If you were there, you may feel uneasy because the N on your compass would point south, and S to the north.

To explain the auroral belt, it is necessary to introduce the concept of geomagnetic latitude. This idea is simpler than it sounds.

The "usual latitude", which is based on the two poles as identified from axle point of the Earth's rotation, is called the geographic latitude.

The geomagnetic latitude is calculated by drawing lines of latitude to join the north and south geomagnetic poles. These lines are drawn at some interval from each geomagnetic pole (see Fig. 4-2).

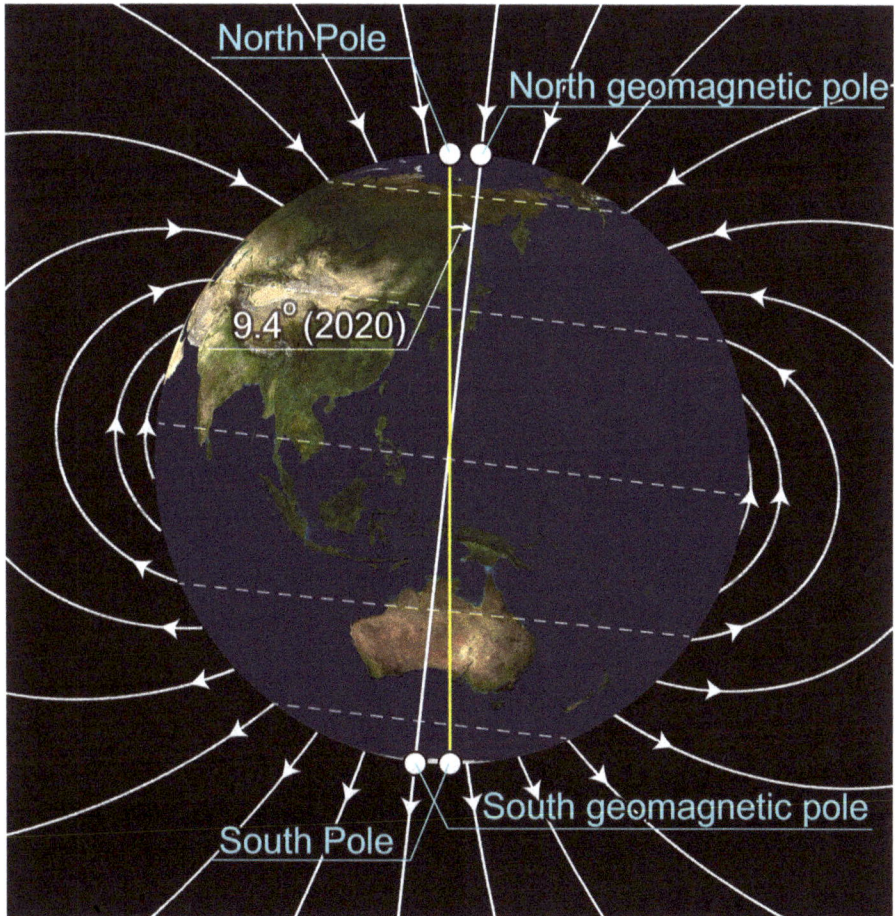

▲ Fig. 4-2. This diagram shows how geomagnetic latitude relates to geographic latitude. Note that they are not identical.

The geomagnetic poles are not the same as the magnetic north and south poles, which compass needles point to. The two geomagnetic poles are located in the northern and southern hemispheres, respectively. The geomagnetic poles are identified by imagining a bar-shaped magnet inside the Earth and calculating the magnetic fields.

Once the latitude lines from the geomagnetic poles are drawn, the two different systems—geographic and geomagnetic latitude—can be compared.

For example, the latitude of geomagnetic coordinates at the east coast of North America is higher by about 10 degrees than that of its geographic coordinates. New York is located at 41 degrees, while its geomagnetic latitude is 54 degrees. The reason for this is quite simple. The geomagnetic pole has shifted toward the East Coast at present. This means that people in that region can enjoy auroral displays without traveling far north. On the other hand, the northernmost place in Japan is located at 46 degrees, while its geomagnetic latitude is only 38 degrees. Between Europe and the west coast of North America, the two latitudes are nearly the same.

The northern auroral belt, where the probability of seeing auroras is highest, is located somewhere at 65–70 degrees in geomagnetic latitude. Now earlier, we calculated how far Italy or Japan is located from the northern auroral belt. Using the probability shown in Fig. 4-1, we can then deduce that, in south Europe or mid-Asia, like in Japan, for example, where the contour line labeled "0.1 of a year" passes through the north of these regions; the probability of an aurora coming down to that latitude in that region in one calendar year is 0.1— or as statistics tells us, about once every ten years.

Note, however, that the statistics shown in Fig. 4-1 were based on observations made by the naked eye, and that no distinction was made between the various types of auroras. For example, whether a faint short-lived aurora near the northern horizon or a very active all-night discrete aurora appears, the occurrence is counted as a day on which an aurora is sighted. The probability of seeing a bright auroral display expanding toward the pole—a so-called auroral breakup (see Sec. 4.10 for more info)—may be less than 0.3; or about 100 days of a year at most. Needless to say, these numbers vary depending on solar activity.

According to more recent high-sensitivity optical observations, auroras are seen in northern Japan several times a year. Even so, these auroras are quite different from the gorgeous colorful auroras shown in Chapters 1 and 2.

4.3 How High Do Auroras Dance?

Contrary to what ancient people thought, auroras are not shimmering in the heavens where the gods live. Similarly, Aristotle's suggestion that auroras were fire coming out from a chasm and Benjamin Franklin's idea that auroras were shining ice in the air of the polar region were also wrong.

Currently, whether or not a proposed theory seems reasonable can be judged by measuring the height of auroras above the ground. A number of scientists spent a lot of time figuring out the altitude of auroras. It was very difficult to estimate the altitude where auroras exist without high-quality cameras.

Carl Størmer, a mathematician and astrophysicist from Norway, was one of the first to challenge this problem. It was the beginning of the 20th century when he took photographs of the same aurora from two ground locations, and using the principle of triangulation, calculated the height of its base from the ground (see Størmer's triangle configuration in Fig. 4-3).

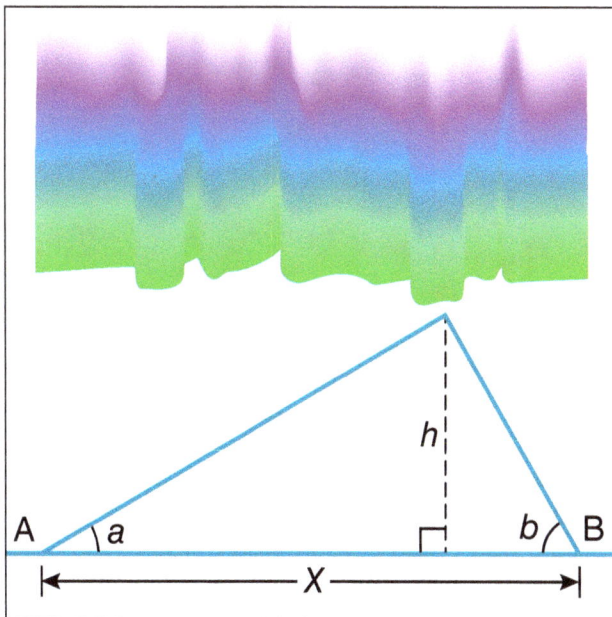

Fig. 4-3. The configuration of key angles needed to estimate the height of the lowest part of auroras from the Earth's surface using the so-called triangle method.

For this purpose, it was assumed that stars in the background of the two aurora photographs were at infinity. The auroras from two different places on the ground are shifted when overlaying the two photographs with respect to the stars. This shift, along with the known distance between the two points on the ground, allows us to calculate the height of the aurora above ground.

The distance of the two ground points should be in the same order of magnitude as the altitude of the aurora to minimize triangulation calculation error. In other words, the triangle should have angles of similar size and sides of similar lengths. At the beginning of this trial, people thought that the bases of auroras began only several kilometers above ground height. This assumption introduced larger errors and resulted in unrealistic calculation values.

Applying this method to approximately 40,000 cases, Størmer obtained the distribution of the height of auroras above the ground (see Fig. 4-4). According to this diagram, the altitude where most auroras are generated is estimated to be near 110 km. It also determined that auroras would never come down to below 80 km in altitude.

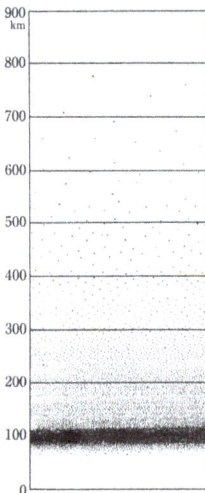

Fig. 4-4. Height distribution of auroras obtained by Carl Størmer, who applied the triangle method to 40,000 individual auroral arcs.

Next, we will talk about the height of auroras above ground by discussing air temperature at different altitudes.

Figure 4-5 shows an average altitude distribution of air temperature. For example, on a jet airplane traveling from San Francisco to Washington, D.C, the pilot would normally announce the "present" flying condition to the passengers in the following way: "We are presently flying over Denver. The altitude is 30,000 ft (or 10,000 m). The outside temperature is –50°F (or –45°C).

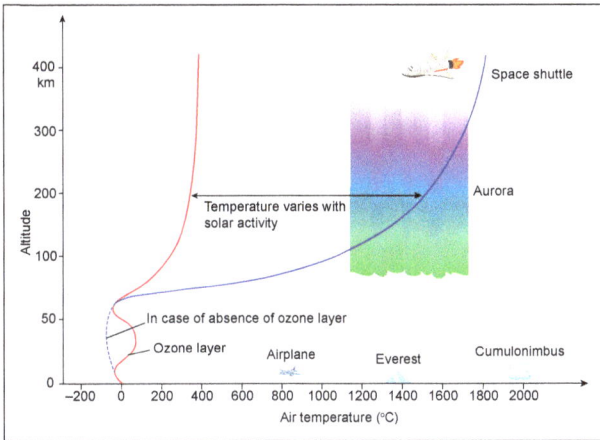

◀ Fig. 4-5. Average altitude distribution of air temperature.

Many people have seen auroras through the windows of airplanes, particularly of international flights in the polar routes (see Fig. 4-6). They claim that auroras are seen at the same altitude level as airplanes in flight. Thus, it is not surprising for them to say that the airplanes fly into auroral curtains. However, this is a false perception of altitude resulting from the fact that the Earth is round, as is evident in Fig. 4-7.

▲ Fig. 4-6. Auroras seen from airplanes. Credit: Shutterstock / Kuznetsova Julia.

▲ Fig. 4-7. When you look at an aurora from the window of an airplane, you feel as if you are flying at the same height as the aurora. This is, however, wrong; it only appears so because of the curvature of the Earth.

Active auroras appear at altitudes between 100 km and 500 km above the surface of the Earth, as discovered by Størmer. That means that the lowest border of auroras is more than ten times higher than Mount Everest, and 3–4 times higher than where the ozone layer is. This means auroras appear in the ionosphere and thermosphere.So, what is the temperature of the altitudes at which auroras appear?

We all know that as we go up higher and higher, the air temperature goes down in conjunction with a decrease in atmospheric pressure. This effect is called the adiabatic process. The adiabatic process is a predictable expansion and cooling of air as it moves up, as well as a predictable contraction and warming of air as it moves down. This effect, however, does not continue forever.

How about the air temperature at auroral altitudes then? If we were to travel inside auroras on the International Space Station (ISS), a hypothetical captain would similarly announce: "The altitude is 400 km. The outside air temperature is 1,000°C." You may not believe it, but this is true. As we go higher up, we reduce our distance from the Sun, and so the temperature rises. In a sense, we are living within the solar atmosphere, as will be discussed in Chapter 5.

4.4 How Bright Are Auroras?

In daily life, our description of light is different from its scientific description. When people say, "It is bright," it means that either the light source is bright, or the luminosity of the object is intense. Scientifically speaking, "brightness" is the degree of intensity of the light source, whereas "illumination" refers to the amount of light our eyes receive.

Let us look at examples of light measurement for auroras. The Rayleigh (R) is a unit of photon flux, used to measure faint light emitted in the sky, such as airglow and auroras. The intensity of auroras can be measured by photons/area/time emitted from the light source.

- The night sky has an intensity of about 250 R.

- $1 \text{ R} = 10^6 \text{ photons/cm}^2$ (column) sec.

- 1 kR (kilo Rayleigh) = 1,000 R.

As a guide, the diffuse aurora is typically 0.5 kR or less. A bright discrete aurora, such as an intense curtain aurora, is 10–100 kR. This means that from a very active aurora, more than 10,000,000,000 photons/cm² are emitted every second.

How bright is an active aurora on the Earth's surface? It is about 0.01–0.1 lux (lx), which is almost the same amount of illumination as reading a book under the full moon. The definition of 1 lx is a measure of the intensity of brightness (or darkness) equivalent to standing 1 meter away from a candle that measures 2 cm in diameter. Again, our experience of light is slightly different from its scientific description.

In reality, how a person senses illumination depends on how large his or her pupil is open. If one has waited a long time under the dark sky for auroras to appear, one's pupils are likely to be more dilated than that of someone who has just emerged from a well-lit environment, and will thus be able to catch more details than the new arrival.

The situation with a camera is a little different. Using a camera with an F1.2 lens and ASA400 film to shoot auroras requires at least a few seconds of exposure time.

4.5 How Are the Colors of an Aurora Determined?

Let us compare the spectrum of sunlight to that of typical auroras. Figure 4-8 shows a depiction of the solar spectrum and the auroral spectrum. The rainbow-like color distribution on the top of the figure represents the solar spectrum. As may be observed, this spectrum is continuous, from red to purple, while the one below it, which represents the auroral spectrum, is not. Instead, the auroral spectrum consists of a number of lines and bands for different colors, which result from energy emissions by atoms and molecules.

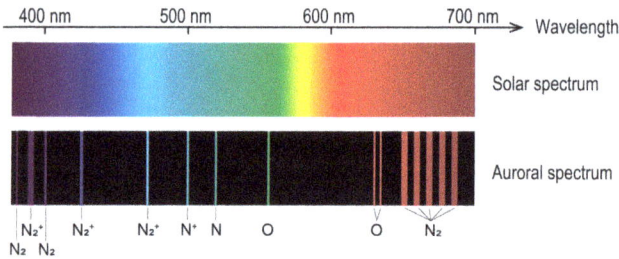

◀ Fig. 4-8. Spectra of visible sunlight and auroras.

Such lines and bands of auroral light can be generated in a vacuum tube when a high voltage is applied between the two ends of the tube. The lights come from the atoms and molecules in the tube when high-speed electrons emitted from the negative electrode, which is connected to a high voltage, strike them.

This process is quite popular in neon signs in cities. The energy state of neon atoms in the tube are changed when hit by streaming electrons. This new state is called the state of "excitation". However, since neon atoms cannot stay excited for long, they must return to the ground state. The familiar red light in a neon-sign is produced and released during this "returning" process.

Here is a description of the discharge tube method of studying how different colors are emitted.

1. By applying a high voltage such as 10,000 volts (V) between the two ends of the tubes, electrons are emitted and accelerated from the end with a negative charge, and they move to the other end.

2. These electrons collide with the atoms and molecules that have remained in the tube, and, gaining energy from the collisions, they shift to higher states of energy.

3. Since the excited state is unstable, the electrons tend to return to their original state. During this returning process, the extra energy is emitted as light.

In the case of auroras, similar processes occur. Auroras are generated from collisions between incoming charged particles, such as electrons and atoms. The electrons originate in the solar corona and solar wind, and the atoms are molecules in the upper atmosphere. The energy of the incoming particle beam would determine how deeply into the atmosphere the beam will penetrate and at what altitudes the auroral light will be produced.

The colors of auroras then depend on "what collides with what"—i.e., they depend on the energy of incoming electrons and the type of atoms and molecules they hit. The principle of quantum mechanics comes into play here.

The whitish green light of the wavelength of 557.7 nm, which is the most commonly seen light of auroras, is emitted by excited oxygen atoms, as shown in Fig. 4-9. Oxygen atoms excited to states above the ground are not happy atoms; like sleepy dogs suddenly jolted awake, they would very much prefer to go back to their stable original states.

Let us account for the most popular process of green and red lights in the aurora by following Davis' (1992) intuitive way of explaining the colors of auroras. Figure 4-9 outlines the process. Initially, the oxygen atom is in what is called the ground state, a condition of minimum energy. When an incoming auroral primary electron strikes an oxygen atom, its impact ejects an electron from the oxygen atom. This creates an excited state, and the green light is the brightest single emission in the aurora.

As Figure 4-9 also shows, the atom has two possible routes to return to its ground state. However, quantum mechanics tends to discourage the "direct" route of emitting 297.2 nm light. One possible route for oxygen atoms to move from the second to the first level is by emitting 557.7 nm (green) light. The oxygen atom can also complete its trip to the ground state by emitting 630.0 nm (red) light.

The oxygen atoms' journey, however, is not always like what Fig. 4-9 indicates, which makes the whole picture a bit complicated, as the lifetime of atoms must be considered. For the sake of simplicity, we will skip discussions along this line, just for this book. Readers who want to know more may consult more advanced publications.

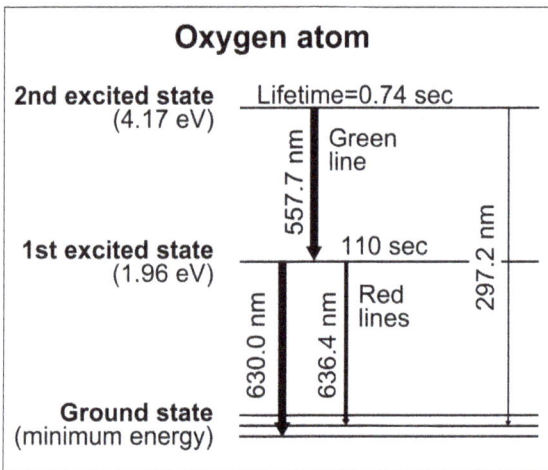

◀ Fig. 4-9. Auroral emission from oxygen atoms, providing the 557.7 nm green line and the 630.0 nm red line. "eV" stands for electron volt, which is a unit of energy.

4.6 How Are Auroral Particles Accelerated in Space?

Although we know that particles are accelerated along magnetic field lines hitting atoms and molecules in the upper atmosphere, we do not know exactly where this occurs. The point of origin of the particles is the Sun, but the route that these particles take and where the acceleration process happens are not yet well understood.

Old popular science books state that auroras are created by charged particles from the Sun that have been pulled in by the Earth's magnetic field, entering the upper atmosphere at high latitudes. Even the famous book *A Planet Called Earth* by George Gamow puts forth this theory. If this simple explanation is correct, auroras would appear mostly on the "dayside" of the Earth. However, this does not explain why "nightside" auroras are much stronger than those on the "dayside".

Another issue is: where and how are auroral particles accelerated? Although all we need is an electric potential difference between higher and lower altitudes, satellites flying over discrete auroras repeatedly observe electrons with energies of keV (kilo-electronvolt, 1.6×10^{-16} Joule). The energy of solar wind particles is not as high as keV, implying that there must be a potential difference of keV along magnetic field lines above discrete auroral forms. One can expect that there must be a "potential V-shaped structure", which accelerates electrons downward toward the Earth, and this same potential structure would accelerate ions in the ionosphere upward. Satellites have, in fact, detected such ions.

4.7 Why Do Auroras Move?

The interesting thing is that auroras do not actually move. What the human observer is actually experiencing is an impression that the lights are moving although the light source is actually stationary. This process is similar to how humans may perceive the light of a neon sign in the city to be moving, when the sign's bulbs are actually completely stationary. A similar effect prevails in the case of auroras.

The seeming motion of auroras is unique. Proper words cannot describe the movement of auroras. At times auroras look like "curtains", and their motions are like that of curtains swaying in the wind. The most interesting and mysterious moment is when an auroral arc starts to move suddenly, and then spread in all directions. We now know that such sudden brightening and expansion occur at the expansion onset of an auroral substorm (see Sec. 4.10 for the definition of substorms).

4.8 Why Do Auroras Look Like Curtains?

People have observed the following two characteristics of auroras that have been likened to attributes of a piece of cloth: they often have "pleats" in the vertical direction, and the bottom edge of the cloth is sharp and clear. A curtain normally hangs vertically. However, auroras do not follow the same rules.

The direction of the pleats indicates the direction of the Earth's magnetic field lines. Although people describe auroras as curtains, these curtains of light are not actually hanging from the top of the sky. The law of gravity does not affect them in the same way it does an actual curtain. They just appear that way to us.

Auroras in mid-latitudes are not vertical. If auroras appear in the equatorial region, the pleats must be parallel to the surface of the Earth. This has been confirmed by artificial aurora experiments conducted in India. In the experiments, a metal vapor cloud was released from a rocket. After being ionized by sunlight, the clouds move under the influence of the Earth's magnetic field lines.

Auroral curtains seem quite simple when viewed from a distance. However, the shape of auroral arcs becomes complicated the closer one approaches them (see the schematic illustration in Fig. 4-10). The colors of auroras vary significantly with altitude. When a corona aurora is directly above us, we can enjoy the apparent motion of different colors, as if they are dancing, and feel that we are being showered by light.

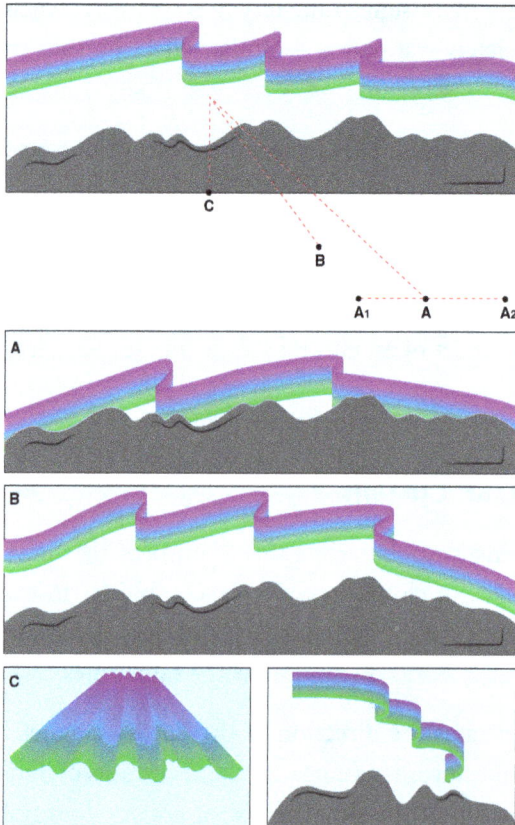

Fig. 4-10. The forms of auroral arcs vary significantly, from simple to complicated, depending on the distance and angles at which they are viewed.

4.9 Are the Auroras in the Northern and Southern Hemispheres the Same?

Sir Joseph Banks, who accompanied Captain Cook, an adventurer from Europe, reported seeing auroras in the southern Pacific Ocean in 1770[1]. By 1800, scientists noticed that auroras in the northern and southern hemispheres were either similar or mirror images. Two points, one in the northern hemisphere and the other in the southern hemisphere, were called conjugate points. The two "conjugacy" points are tied by a field line.

One of the best pairs of conjugate points is between Iceland and the Syowa Station in Antarctica, or between Fairbanks, Alaska and Macquarie Island, a UNESCO World heritage site 1,500 km southwest of Christchurch, New Zealand. At middle latitudes, Rikubetsu in Hokkaido, Japan, and the south of Melbourne, Australia, constitute a unique pair. A research team from the University of Alaska and the Los Alamos Scientific Laboratory conducted a total of 18 paired experiments using aircraft. A pair of photos came from these experiments, showing an example of exact conjugacy. Some dissimilarities do exist between two conjugate points, however.

Figure 4-11 shows a photograph by Polar, which was near the equator observing the auroral ovals in the two hemispheres at the time. It is noticeable that the entire auroral ovals are quite similar, in that the bright points in the two hemispheres constitute a pair, and the dark areas in the two hemispheres seem connected by a field line.

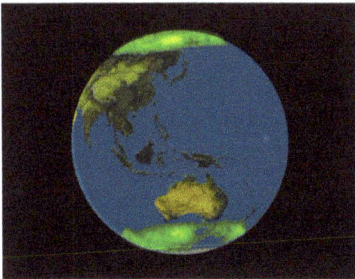

◄ Fig. 4-11. Simultaneous distribution of auroras in the northern and southern hemispheres, seen from the Polar satellite. Credit: NASA.

1 "At about 10 o'clock a phenomenon appeared in the heavens, in many things resembling the aurora borealis, but differing materially in others: It consisted of a dull reddish light reaching in height about twenty degrees above the horizon." (Banks, J., Journal of the Right Hon. Sir Joseph Banks, Cambridge University Press, 1896).

There is no guarantee, however, that the two points are connected by a field line at all times. In fact, when auroras are very active, electrical currents flowing along the field line can destroy the pairing of stations.

4.10 What Is an Auroral Substorm?

When we see auroral displays from the ground, it seems like that they occur at random, showing very complicated shapes along with many local structures. When one sees them "dancing in the sky", it seems like that there is no systematic way that auroras come and go. But is this really the case? It was Akasofu and Chapman (1961) and Akasofu (1964) who proposed a framework of an auroral substorm to systematically explain all complicated structures of auroras.

Using a number of all-sky camera records acquired during the International Geophysical Year of 1957–58, they tried to account for local behavior seen in all-sky camera data. They looked at the systematic growth and decay of auroras, otherwise known as an auroral substorm. "Sub", in this case, means elementary or basic, not secondary. In this case, a "substorm" is quite similar to the elementary storm proposed by Birkeland (1908).

Figure 4-12 shows a schematic illustration of auroral distribution and its dynamic changes in geomagnetic latitude and local time as a function of substorm time. Developed by Akasofu (1964) as a result of morphology using a number of all-sky camera observations, this first-generation model of substorms has been improved over the years to account for complicated individual substorms, especially substorms observed by various satellites.

In this context, T = 0 indicates the time at which a substorm begins with the so-called breakup. What comes before the breakup is a slow equatorward motion of quiet-time aurora. Within 5–10 minutes after the expansion onset, i.e., breakup, an auroral bulge is formed, rapidly moving poleward as well as westward. At the western end of the bulge, there is a unique large-scale structure called the westward traveling surge (WTS). Half an hour from the WTS expansion onset, the surge reaches the highest latitude in the midnight meridian, although it continues to move westward.

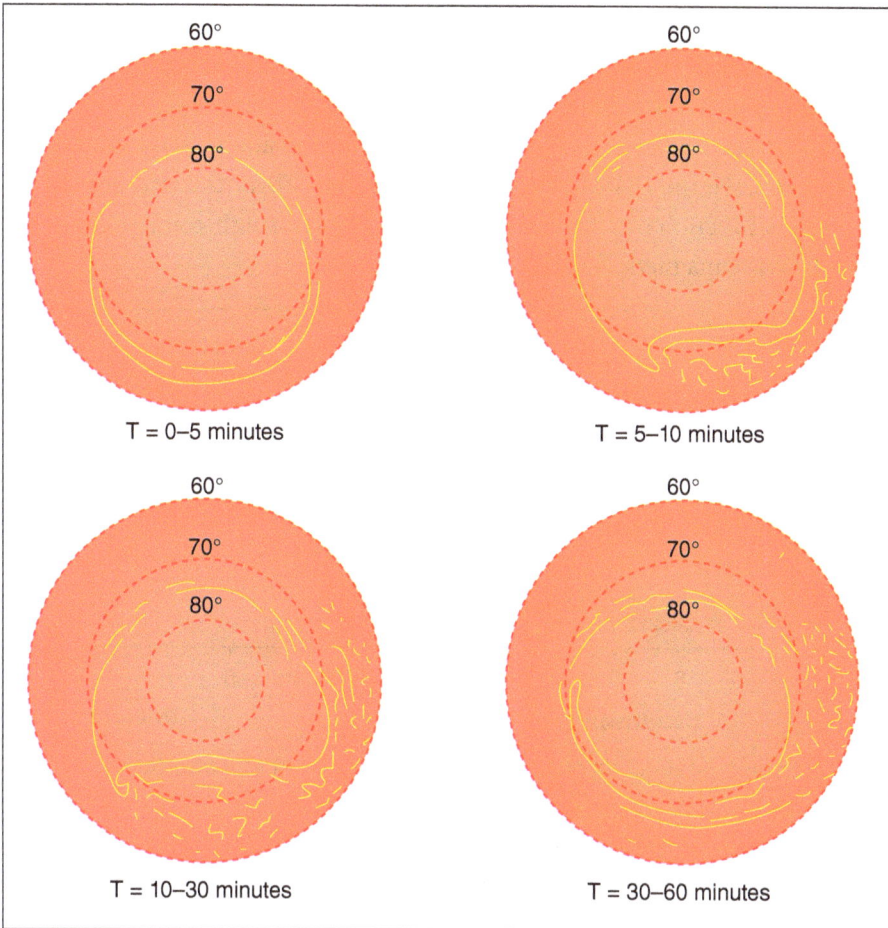

▲ Fig. 4-12. Auroral distribution during an auroral substorm. T = 0 is the so-called expansion onset. Adapted from Akasofu (1964).

4.11 What Determines the Probability of Auroras?

In extensive studies using solar wind data, it has been found that substorm intensity and the probability of substorm occurrence are both heavily controlled by the northward or southward oriented polarity. An auroral breakup is a unique phenomenon, signaling the sudden beginning of a magnetospheric substorm: a substorm is a manifestation of energy storage and the subsequent release in the Earth's magnetosphere and ionosphere, as well as the inner magnetosphere.

Figure 4-13 shows the probability of substorm occurrence, given as a function of the north-south component of the interplanetary magnetic fields. Here, the electron spectra observed by ISIS-2 have been used to calculate the substorm's probability. First, the spectra observations are evaluated to see if they have the characteristics of a substorm. Then, the relevant data is shown as a function of 1-hour averages of the north-south component of the interplanetary magnetic field.

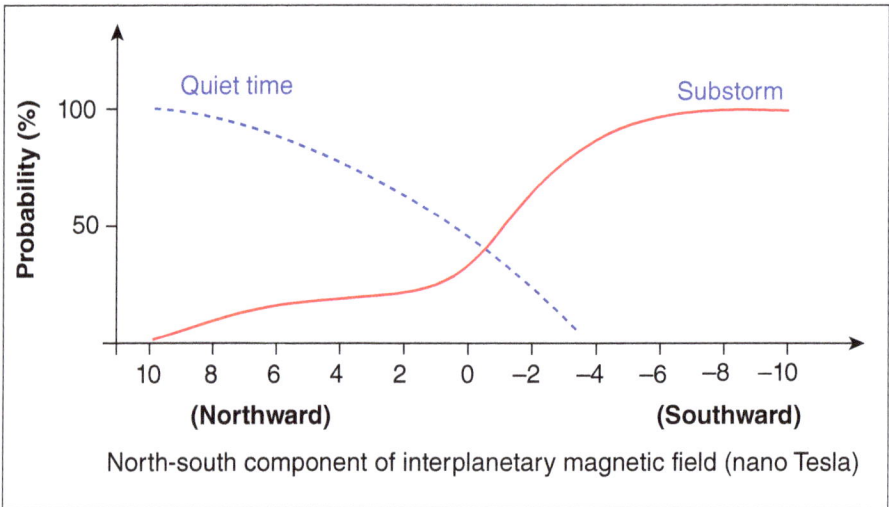

▲ Fig. 4-13. Probability of substorm occurrence, given as a function of the north-south component of the interplanetary magnetic field. Adapted from Kamide et al. (1977).

The implications of the diagrams are as follows. When the interplanetary magnetic field (IMF) is directed southward, and plasma convection gets enhanced in the magnetosphere, interactions between magnetotail and polar processes are caused, enhancing the probability that polar auroras could become quite active. The suggestion that this reconnection between the southward IMF and northward geomagnetic field plays a dominant role in transferring solar wind energy has since been confirmed.

4.12   What Are Auroral Electrojets?

Electrical conductivity is relatively high inside auroras, and the electrical currents flowing inside are caused by the interactions between the solar wind and the Earth's magnetic field. Because these electric currents flow in a rather spatially limited area, they are called auroral electrojets. Their intensity in the aurora is enormous, measuring as much as 10^7 A.

How do we measure the intensity of auroral electrojets? Figure 4-14 shows the average distribution of electrical currents outside the aurora as well as the auroral electrojets. The spatial distribution of Joule heating from the ionospheric currents is also shown. Ground magnetic disturbances are also required. There are some 300 observatories on the Earth's surface to measure the distribution of ground magnetic disturbances. By using the ground magnetic inversion techniques, it is possible to compute the distribution of the auroral electrojets (Kamide et al., 1981).

From Fig. 4-14 we learn that auroral current consists of eastward and westward auroral electrojets in the afternoon and evening sector, and midnight and morning sectors, respectively.

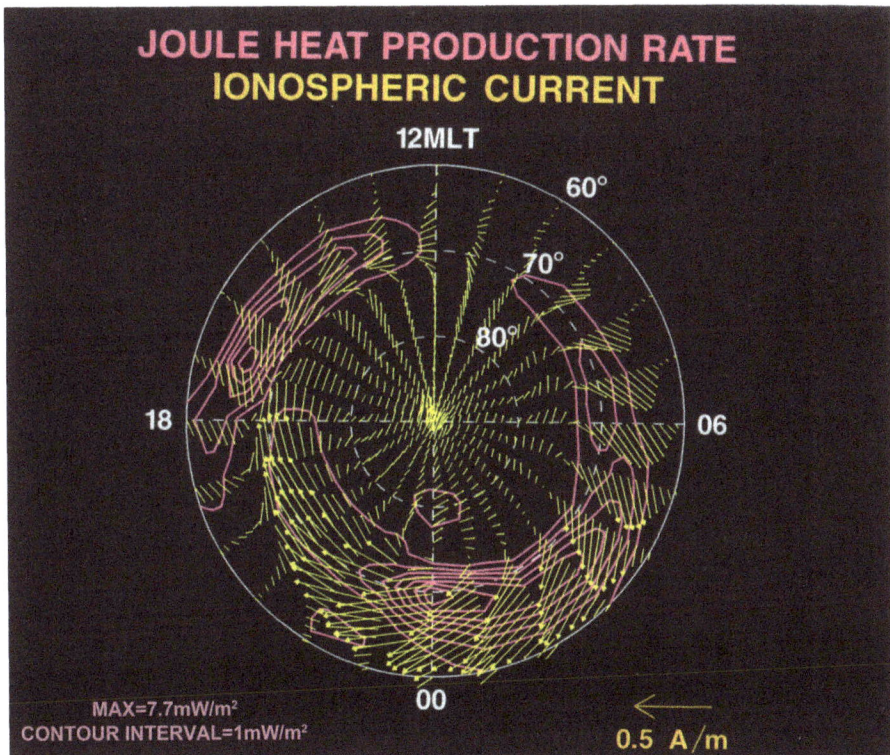

▲ Fig. 4-14. Average distribution of electrical currents outside the aurora as well as the auroral electrojets. The spatial distribution of Joule heating from the ionospheric currents is also shown by the reddish contour lines. The geomagnetic north pole is located at the center, and the Sun is to the top. MLT stands for magnetic local time.

References

- Akasofu, S.-I. (1964), The development of the auroral substorm, Planet. Space Sci., 12, 4, 273–282, doi:10.1016/0032-0633(64)90151-5.

- Akasofu, S.-I., and S. Chapman (1961), The ring current, geomagnetic disturbance, and the Van Allen radiation belts, J. Geophys. Res., 66(5), 1321–1350, doi:10.1029/JZ066i005p01321.

- Birkeland, Kr. (1908), The Norwegian Aurora Polaris Expedition 1902-1903, Vol. I. On the cause of magnetic storms and the origin of terrestrial magnetism (First Section). Christiania, H. Aschehough & Co., 1–315.

- Davis, N. (1992), The Aurora Watcher's Handbook, University of Alaska Press, ISBN:978-0912006598.

- Kamide, Y., A. D. Richmond, and S. Matsushita, (1981), Estimation of ionospheric electric fields, ionospheric currents, and field-aligned currents from ground magnetic records, J. Geophys. Res., 86(2), 801–813, doi:10.1029/JA086iA02p00801.

- Kamide, Y., P. D. Perreault, S. -I. Akasofu and J. D. Winningham (1977), Dependence of substorm occurrence probability on the interplanetary magnetic field and on the size of the auroral oval, J. Geophys. Res., 82(35), 5521–5528, doi:10.1029/JA082i035p05521.

Chapter 5

Implications of the Delicate Balance Between the Sun and the Earth

As we learn more about aurora science, it becomes evident that auroras are not just mysterious objects in the sky.

An aurora almost always moves when it is active and bright. Viewers describe auroras as "dancing on the big stage in the night sky", while scientists say that auroras signify a unique manifestation of the relationship between the Sun and the Earth, which is essential to life on the Earth. The former, which governs the entire solar system, is restlessly sending us important messages through auroral displays in the polar sky. How should we interpret the Sun's messages through the auroras seen on the Earth?

One may wonder: "If the Sun really causes an aurora, then why do auroras appear more actively on the "nightside" than on the "dayside" of the Earth? Even scientists who study auroras are still unable to answer this simple question easily. Instead, new questions arise whenever a new type of satellite data becomes available. These new questions help to guide us in understanding the complex relationship between the Earth and the Sun.

In recent years significant progress in auroral research has resulted in considerable development and innovation in satellite and radar techniques. This progress has unveiled a series of discoveries about the solar-terrestrial relationships involved in the creation of auroras. For example, we have unveiled the mechanism by which charged particles come from the Sun toward the Earth. These charged particles cause auroras in the night sky. There are incredibly complicated phenomena occurring in space between the Sun and the Earth. These phenomena and auroras are closely and intricately related with each other. They are an essential part of enabling all life on Earth.

5.1 At the Border Between the Universe and the Earth

Many people regard auroras as "beautifully-dancing lights" in the polar sky, though few might realize that auroras are generated at the border between space and the Earth. The fact that auroras exist at the border of space is not a coincidence—the conditions of both regions overlap where auroras manifest themselves.

The height at which auroras occur extends from about 90 km to a few hundred kilometers above the Earth's surface, reaching the upper atmosphere, which is the boundary between space and the Earth. One can say that the "wavy curtains" of auroras express what is happening in space. However, what is the practical difference between the Earth and space? What, and how, do we divide these two entities?

To answer this question, we suggest looking around—the materials that surround us typically exist in one of these three states: solid, liquid and gas. We know that energy and heat are involved in changing an object's existing state to another state. When ice (i.e., H_2O in solid form) is heated, for example, it turns into water (i.e., the liquid form of H_2O). Water boils at 100°C and changes into water vapor (i.e., the gas form of H_2O). The critical temperature key to changing the states depends on each material. For example, iron and copper turn into high-density liquids at 1,000°C, which then transform into gas forms at 2,000–3,000°C. We also know that mercury, which is used in clinical thermometers, is a rare metal because it maintains its liquid form at temperatures between −39°C and 357°C.

What happens to a material when it is heated? How does heat energy travel inside the material? To answer these questions appropriately, we need to talk about the materials that make up our Universe and the Earth from a molecular activity point of view.

The higher one climbs a mountain, the lower the temperature becomes. It is the same the higher we ascend into the sky, up to around 10 km above the Earth's surface, where the temperature of the Earth's atmosphere reaches its lowest. This is also the altitude at which many commercial airplanes fly. The temperature begins to rise above the altitude of 10 km and continue to do so even up to about 100 km—the level where auroras are typically at their brightest. The reason for this increase in temperature is because we are now

▲ Fig. 5-1. An auroral display through a window of the International Space Station. It is a curtain of light shining at the boundary between space and the Earth. This picture was taken by the crew of Expedition 30 of the International Space Station on March 4, 2012. Credit: NASA.

closer to the Sun. In fact, the temperature at a height of 200 km is more than 1,000°C. At this altitude, temperature changes in the range of a few hundred degrees can occur quite easily, depending on solar activity.

Under conditions where the atmosphere is exposed to strong solar energy, molecules cannot keep an electrically neutral state. Normally, all materials on the Earth's surface are in a neutral state. In other words, their positive electricity and negative electricity cancel each other out. At a height of 100 km, where the solar energy is stronger than the attractive force within a molecule, the attractive force between negatively and positively charged particles becomes negligible. This means that a world totally different from ours exists 100 km above the ground. We may call it an electric world consisting of plasmas, in which negative and positive charges are separated, although their total numbers are nearly equal. This plasma state is beyond the states of matter—solid, liquid and gas—we are accustomed to on the Earth.

Plasmas occupy more than 99% of the Universe. Earth is an exception, in the sense that it consists of neutral gases, essential for sustaining life. See Fig. 5-1, taken from the International Space Station (ISS), which shows auroras brightening 100 km above the Earth's surface.

5.2 The Earth Resides in the Sun's atmosphere?

You may be surprised if you hear from scientists that the Earth is located inside the Sun's atmosphere. The solar wind coming from the solar corona, from which auroras are born, is indeed part of the Sun's atmosphere. This high-temperature solar gas is called the corona. The corona gas escaping from the enormously strong solar gravitation in all directions becomes the solar wind.

Since our Earth has a magnetic field, life is protected from this solar gas. In other words, the Earth is completely covered by the solar gas and receives every breath the Sun takes in the form of dancing auroras. Auroras are an image of the Sun's breathing, projected on a giant screen of the upper atmosphere in the polar sky.

The Sun is far from a perfect and stable sphere, unlike what ancient people envisioned. Many believed the Sun to be a god. However, the Sun keeps changing dynamically and is not the generous protector that many have thought it is.

Many exist on Earth without realizing that we are being protected from the harsh environment in space by two barriers: the very thin veil of the atmosphere and the Earth's magnetic field. The former protects all life from harmful ultraviolet rays and x-rays. The latter guards us from plasmas which fill up the space between the Sun and where the Earth's atmosphere ends, as well as from high-energy cosmic rays scattered in space. If the Earth had no magnetic field, solar wind (i.e., the energy flow of plasmas from the Sun) would directly hit the Earth and even blow off its atmosphere.

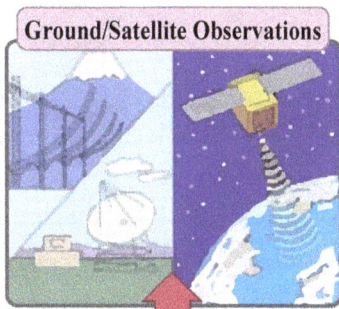

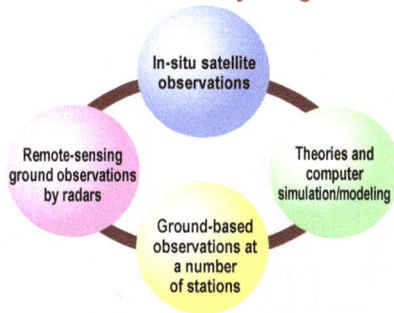

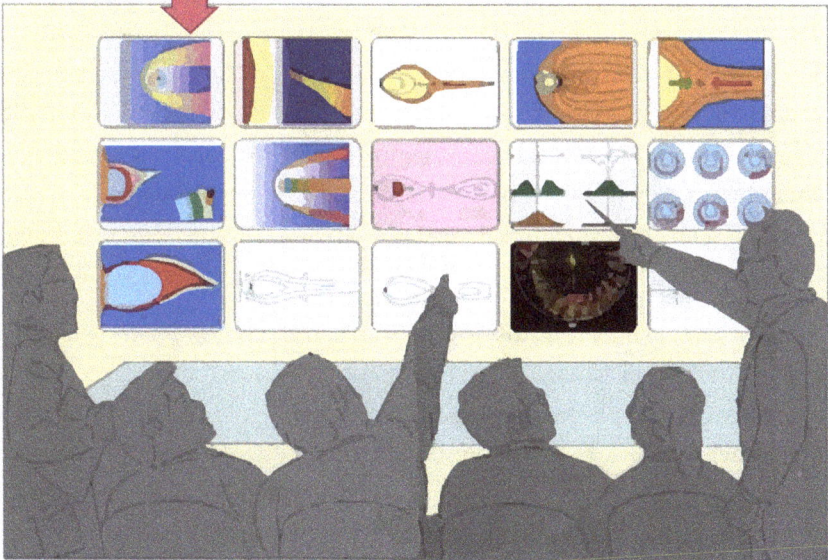

▲ Fig. 5-2. A unique laboratory in the interplanetary space between the Sun and the Earth. In this giant experiment room, scientists can directly test new theories against data obtained from spacecraft observations. Knowledge acquired through studies of plasma processes occurring around the Earth, such as auroras, is applied to other scientific fields, e.g., nuclear fusion. The GEDAS (Geospace Environment Data Analysis System) has been used to study and acquire new knowledge on plasma processes occurring around the Earth.

Since we are well protected by these two barriers, we hardly notice what happens in space. Auroras, on the other hand, tell us of the electromagnetic processes occurring in the interplanetary space between the Sun and the Earth. They also enable us to better understand the Sun–Earth relationship by projecting the space phenomena onto a "screen" that is the Earth's atmosphere.

All life on the Earth and auroras have common sources: the atmosphere and the Earth's magnetic field. Thus, it can be said that auroras are proof that we can survive on this planet.

5.3 Natural Laboratory in Space

You might think that aurora scientists study auroras every day. However, in reality, no researcher in the world works solely on auroras. Though the auroral phenomenon relates closely to various phenomena that occur in the heliosphere and the upper atmosphere of the Earth, auroras are not treated as an independent phenomenon; but rather, as a study in trying to understand the Sun–Earth relationship.

Toward that goal, the aurora is a clue given to us by nature. The space between the Sun and Earth is a unique laboratory. This natural laboratory provides researchers with opportunities to verify theories and models that they have constructed around the fundamental laws of electromagnetism and plasma physics. For instance, basic questions like "where does auroral energy accumulate" and "how do auroral particles gain acceleration" can be studied in this natural laboratory (see Fig. 5-2).

When a new principle is discovered in plasma physics, its knowledge will be applied to auroral science. Conversely, aurora research may also contribute significantly to basic plasma physics. Plasma physics is fundamental to nuclear fusion, which releases a vast amount of energy and is known to be one of the powerful energy sources of the Sun. In the fusion field, it is essential to keep plasmas in a bottle formed by a specific magnetic configuration. In the case of auroral research, plasmas are naturally held in the magnetotail, which is formed by interactions between the solar wind and the Earth's magnetic field. It is safe to say, in this sense, we learn a lot about the future of our energy issues through auroral research.

5.4 Auroras Sometimes Trick Us

It may be difficult to imagine that auroras, despite their beautiful forms, can sometimes do mischief and disturb human society.

The electrical currents flowing inside auroras can vary dynamically and induce abnormal electric fields in our modern electrical equipment. There can be more than ten million amperes of electrical current flowing inside active auroras. This large fluctuation in auroral electrical currents causes abnormal electric fields in transmission lines, which then damage transformers and circuit breakers.

Auroras have triggered blackouts and serious damage to equipment. It takes only a few minutes to cause costly damage to electrical infrastructure. For example, on March 13, 1989, now called Black Monday, six million people in northeastern Canada were forced to live without power because of an enormous current flow in the aurora that appeared that day.

Auroras can affect other aspects of our modern infrastructure as well. For example, auroral currents can accelerate corrosion in oil pipelines. This corrosion results from an abnormal current induced by auroras, along the pipelines. It has been pointed out that the heating from the abnormal current flow can put the surrounding ecosystem out of order.

The extent of an aurora's mischief is not confined to the power system on the Earth's surface. It also brings about communications disorder. When radio signals were disrupted in England during the Second World War, people blamed the German army but never suspected auroras. In another incident that happened on February 8, 1986, two trains were involved in a head-on collision in the Canadian Rockies. An active aurora was deemed as one of the primary suspects—its intense auroral current disturbed the trains' radio communication signals. This type of event can be very serious in a vast country such as in Canada, which controls train movements using radio communication. Navigation systems for boats and cars can also be disrupted by an auroral current.

Auroras can also emit heat. They are just like an electric hot plate, and the continual heating of the upper atmosphere caused by an aurora changes the atmosphere's composition. The wind patterns are also affected, and low altitude satellites often lose their correct orbit as a result. Their lifetimes are also shortened due to this expansion of the air. High technology instruments on board the satellites also sustain damage when hit by high-energy particles coming from the Sun (see Fig. 5-3).

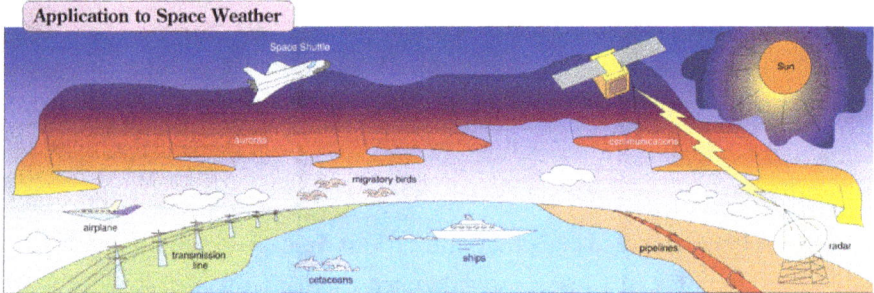

▲ Fig. 5-3. An aurora is a phenomena of space weather that originates from solar activity. Space weather influences the ecology on the Earth as well as satellites, communications networks and power lines, which are indispensable to modern life.

We can no longer dispense with electricity, communications or satellites in this modern life. As the importance of technology in modern society increases and we interact with space more in our daily life, we face increasing exposure to auroral mischief. This growing dependance on tech, however, also ensures that there is continual progress in communications and space development, to make our lives more convenient as well as reduce damages sustained as a result of disasters. For example, an image of clouds seized by satellites enables us to forecast a typhoon's path as well as other weather patterns. An earthquake's epicenter can be determined because of advancements in computer technology. But, when we view this progress from another angle, we immediately notice that civilization has advanced to the point where its vulnerability to the influence of auroral mischief has also increased.

Computers onboard satellites and communication technology are especially affected by auroras and space weather. During the Lillehammer Winter Olympic Games in 1994, satellite broadcasting abruptly halted. This event occurred when a Japanese ski jumper, a candidate for a medal, was competing. The same magnetic storm also caused four satellites in Canada and the USA to malfunction. The resulting confusion among hundreds of newspapers, magazine companies and TV stations was unprecedented.

The effects of auroras on power transmission lines have also increased: as lines become longer to transfer electricity over greater distances, they are more susceptible to the influence of induced electricity from auroras. There have been many efforts to reduce the weight and size of satellite instruments, but that has led to the instruments' increased vulnerability to high-energy cosmic rays.

These "bad weather events" in space not only bring inconvenience to modern civilization but may also deal a blow to the world's economy. It is also not an exaggeration to state that our lives can be endangered because of space weather. Research projects aiming to understand and forecast space weather have already been launched with the view to prevent various disasters and confusion and to minimize the resulting damage. By correctly forecasting electromagnetic weather in the space surrounding the Earth, these disasters can be minimized. In the USA, six departments from the federal government have conducted a 10-year joint project to do just that. Predicting the condition of the upper atmosphere accurately is as important as predicting the occurrence of auroras.

5.5 Auroras and Life on the Earth

How is it possible for pigeons to fly in the right direction even in the dark? This had long been a mystery since the middle of the 19th century when pigeons began to play an important role in communications in Europe. Since the pigeons could fly without getting lost even in bad weather, they were examined to see whether they made use of the Earth's magnetic field as an orientation tool. A number of experiments in the USA and Europe to better understand this "supernatural ability" of pigeons were conducted repeatedly. Results showed that pigeons somehow managed to carefully combine various types of information such as the zenith angle of the Sun and star constellations and use them as a compass.

During these studies, special contact lenses were used on the pigeons, and the following observations were made:

1. Pigeons could not fly in the correct direction when they wore Helmholtz coils, which disturbed the magnetic field; and

2. They took a wrong course when the magnetic field was distorted by high auroral activity.

Thus, it was confirmed that carrier pigeons rely on the Earth's magnetic field to orientate themselves when flying, and that they can get lost when an intense geomagnetic storm is in progress. A notable incident that proved the latter occurred in June 1988, when a magnetic storm hit the International Pigeon Race, and only 5% of the 5,000 participating pigeons found their way from France to their destination in England.

Other animals are also able to perceive the Earth's magnetic field; for example, seagulls and other migratory birds that travel as far as hundreds of kilometers without stopping, and salmon that return yearly to their river of birth. Mammals like dolphins and whales may use the magnetic field for navigation as well. A research team from Los Angeles Harbor College and University of California found that dolphins have magnetic sensors in their cerebrum (Zoeger et al., 1981)—the first such finding in a mammal.

When the magnetic field is disturbed, these mammals are sometimes stranded on the shore in a group. According to recent research conducted at Kiel University, group-strand events occurred primarily at high latitudes during a magnetic storm (Vanselow, 2020).

There is one more example of a living thing that detects the Earth's magnetic field. This organism, which is affected negatively by oxygen, is the magnetic bacteria living in lakes, marshes, shallows and the mud in marshy places. Known as magnetotactic bacteria, they move in lines relative to the magnetic field. In the northern hemisphere, they run into mud seeking food in the same direction as the magnetic field. On the other hand, similar bacteria in the southern hemisphere go in the opposite direction of the magnetic field.

But what about humans? Do the magnetic field disruptions caused by auroras influence human bodies? These are important questions especially in space science and exploration because astronauts in space are constantly exposed to space weather including the effects of solar wind when auroras form.

Some researchers believe that the Earth's magnetic field may have some effect on our emotions as well as the efficiency of our work, and research has been conducted on the relationship between auroras (magnetic storms) and heart failures (Stoupel et al., 2015). We also cannot discount the possibility that the electricity produced in auroras affects our way of thinking and actions, although it is difficult to prove these assumptions because we determine direction mostly by visual cues. But can humans sense the Earth's magnetic field? Is this our "sixth sense"?

A research group at the University of Manchester carried out a series of experiments to determine our ability to detect the magnetic field (Baker, 1980). Research was carried out using student volunteers who wore a bandage over their eyes and electric coils around their heads to "eliminate the effects of the Earth's magnetic field". Results of the experiment suggested that women might be more sensitive to the Earth's magnetic field than men (Baker, 1982).

So, how do we sense the magnetic field? So far, there has been no answer to this question, though some believe the cerebellum, adrenal glands, or other organs are involved.

What do these mean for astronauts then? For instance, in October 1989, an intense magnetic storm occurred and a splendid aurora was observed over Hokkaido, Japan. Given what we know about auroras and particles in solar wind, we now know that what appears as a beautiful light show in Earth, is far less benevolent at auroral altitudes and beyond. Had any astronauts been outside the space shuttle during the event, the high-energy particles generated at the time could have been fatal to them. Astronauts can only work outside the space shuttle when the Sun is quiet; in other words, when auroras are weak.

5.6 Will Auroras Move to Lower Latitudes?

Many believe that auroras are generated under low-temperature conditions since they are seen at high latitudes. Studies over the years show that we live in a time on the Earth when auroras are observed in cold areas. However, the locations of auroras in the past, present and future may be different and are not influenced by the temperature levels at any one location. The latitudinal range over which auroras appear depends on the balance between solar wind and the geomagnetic field.

While it is true that places like, say, Japan are presently located far from the region where auroras are most active, this is because the Earth's geomagnetic axis currently differs from its rotational axis by only 9.4 degrees as of 2020. And this is something which can change. For example, on Neptune, its rotational and geomagnetic axes differ by several tens of degrees. In such a situation, auroras would more likely exist near the planet's equator.

The Sun, the second factor influencing where auroras manifest, is quite dynamic—solar flares, coronal holes and sunspots are actively changing location and intensity. A stream of plasma emanating from the solar corona is called solar wind. When the speed of solar wind moving toward the Earth is 1,000 km/sec, it makes the Earth's magnetosphere smaller. The pressure balance between solar wind and the geomagnetic field determines the location of the auroral belt on Earth. The averaged location of the belt is near 65–70 degrees in geomagnetic latitude. During more intense solar wind speeds, the auroral belt shifts equatorward and auroras can be seen from Hokkaido, Japan.

Full displays of auroras cannot be seen from Hokkaido, though. Only the upper portions of auroral curtains can be seen, since the Earth's surface is sphere-shaped. This top part of auroral curtains is mostly red, as explained in Chapters 1 and 2, and this color can be seen above the horizon from Hokkaido. There is also a possibility that red auroras can be observed from other parts of Japan, as the auroral belt comes down to the central part of the Kamchatka Peninsula during very intense geomagnetic storms as a result of high-speed solar wind.

How often does the auroral belt expand to the Kamchatka Peninsula? Figure 5-4 shows the auroral belt location in March and October of 1989. Based on precipitating electron data obtained from the Defense Meteorological Satellite Program (DMSP) satellites, the location of the auroral belt, in geomagnetic latitude, was closest to the equator during peaks of solar activity.

On March 13, 1989, auroras came far south from their ordinary auroral location, owing to a historically large geomagnetic storm. Satellite orbits went

March 1989

October 1989

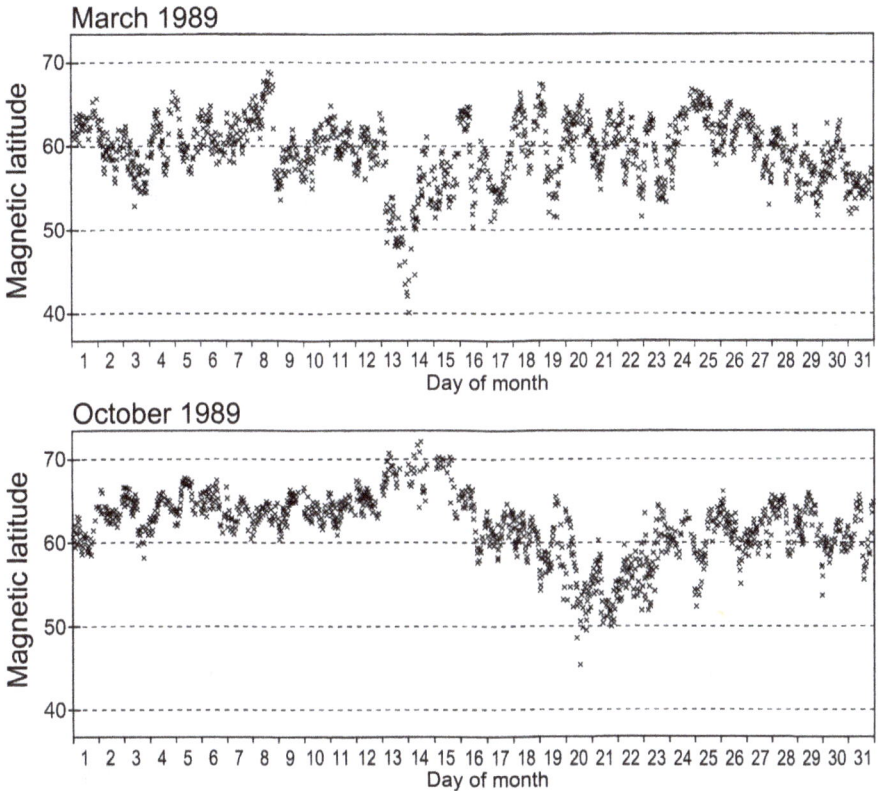

▲ Fig. 5-4. In what latitudinal range does the auroral belt shift? This figure shows the equatorward boundary of auroral precipitation for March and October 1989 based on DMSP satellite observations. Provided by Nobuhiro Yokoyama and Frederick J. Rich.

off course, and in Quebec, Canada, a power outage lasted for hours because induction electric currents generated by large-scale auroras hit its largest power plant. Auroras were reportedly observed even in Florida, which was as low as 40 degrees in geomagnetic latitude.

This storm was enough for auroras to be seen from Japan, but, unfortunately, the auroras returned to their ordinary latitude by nightfall. The second chance arrived on October 21 that same year—a red aurora was observed in Hokkaido, at last (see top of Fig. 5-5).

Auroras have visited Hokkaido several times after that. For instance, on June 5, 1991, a faint picture of a red aurora was taken in Rikubetsu, Hokkaido (see bottom of Fig. 5-5). According to older records, auroras were seen, on average, every ten years in low latitude regions, such as in Japan. However, more recent observations by high-resolution cameras indicate that they have visited Hokkaido more often than that.

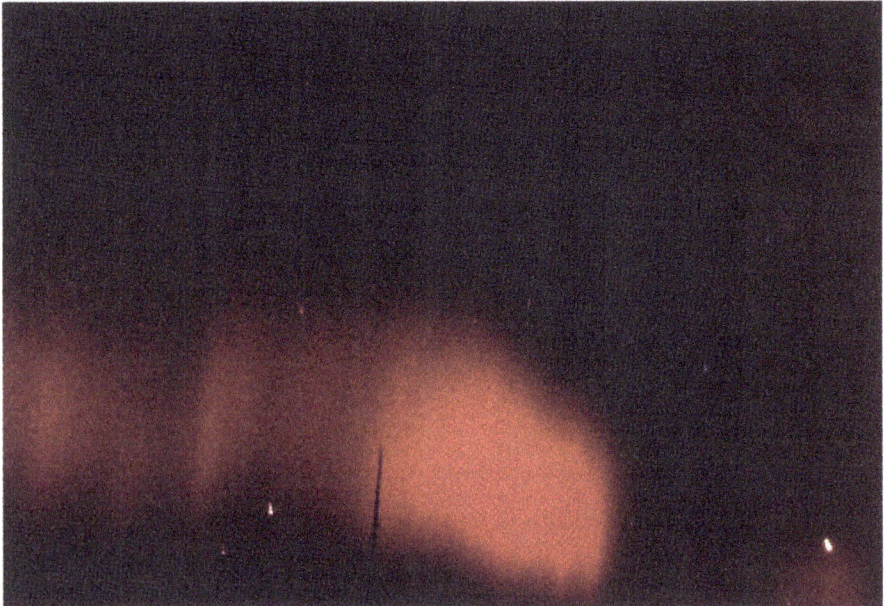

▲ Fig. 5-5. (Top) On October 21, 1989, a red aurora was observed in Rikubetsu, Hokkaido. Photo taken by Takashi Maruyama. (Bottom) A faint aurora visited the same town again on June 5, 1991. Photo taken by Susumu Murata, and provided by Masako Izawa.

The same phenomenon, i.e., seeing auroras at low latitudes, occurs not only under high solar activity but also when the Earth's magnetic field becomes weak. These two processes are equivalent in terms of the pressure balance between solar wind and the geomagnetic field. That is, auroras will appear over Japan when the geomagnetic field weakens, even if solar activity remains unchanged.

The geomagnetic field has indeed been getting weaker since the beginning of the 19th century, when German scientist Carl Friedrich Gauss measured, for the first time, the intensity of the Earth's geomagnetic field. The field has been steadily decreasing at a ratio of ~0.05% a year since then (see Fig. 5-6). According to accurate figures of the geomagnetic field over the past 40 years, the decreasing rate is ~0.06% a year (Alken et al., 2021). At the current rate, our Earth's geomagnetic field will reach zero in 1,600 years!

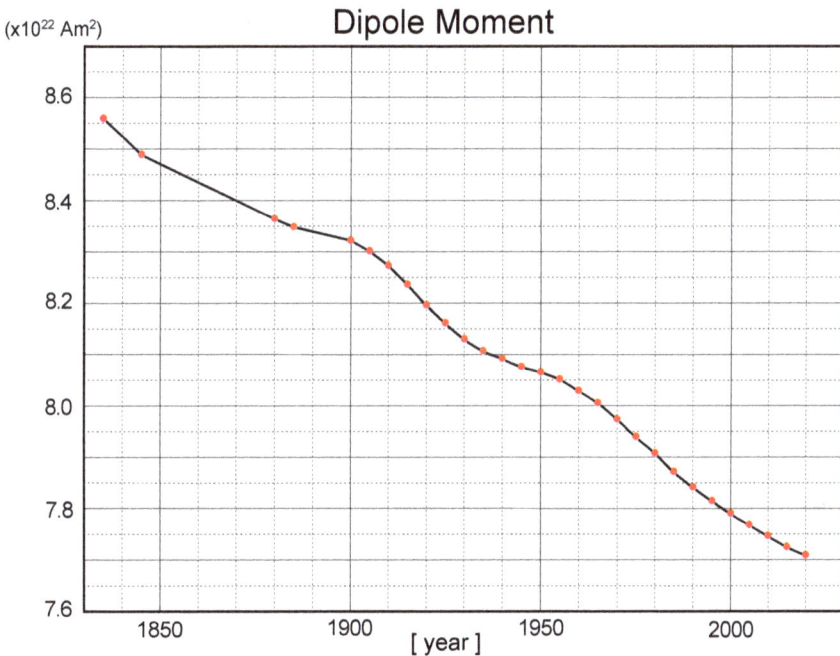

▲ Fig. 5-6. Currently, the Earth's magnetic field is steadily decreasing. In this graph, the magnitude of the Earth's magnetic field in terms of the magnetic moment of the dipole magnetic field is plotted as a function of years. Provided by World Data Center for Geomagnetism, Kyoto.

The auroral belt is predicted to be above Japan a thousand years from now, as shown in Fig. 5-7. The reason why the auroral belt is in the shape of a disordered circle in Fig. 5-7 is that the terrestrial magnetic field does not consist of a pure magnetic dipole, but has complicated multi-order components. The influence of the gradually shifting magnetic poles was also taken into consideration in this prediction.

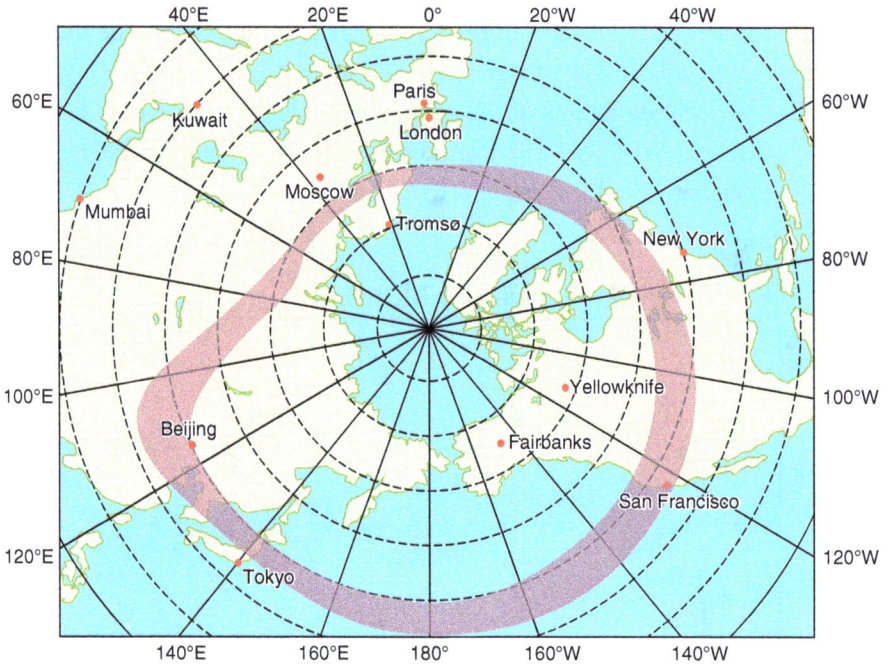

▲ Fig. 5-7. Predicted location of the auroral belt in a thousand years. If the geomagnetic field continues to decrease at the current rate, auroras will appear over Japan at that time. The auroral belt is not depicted as a perfect circle because the Earth's magnetic field is not a pure dipole. Reconstructed from figure provided by Takashi Oguti.

There is an important assumption in this prediction: "If the geomagnetic field keeps decreasing at the present pace". Scientifically, we are not able to predict the exact size of the auroral belt for the next thousand years with the data we have right now. We need at least 1,000 years' worth of data to prove, or disprove the correctness of this figure.

It may be too long to wait. A geomagnetic field that becomes too weak will not be able to shield life on the Earth. The weak geomagnetic field will allow high-energy particles coming from the Sun and galaxy to break through it, and thus easily reach us. This can cause damage to our genetic code, resulting in gene mutation, sickness or even death. In addition, without a geomagnetic field, there would no longer be an atmosphere on Earth; solar wind directly blowing against the Earth would deprive us of oxygen. Thus, one should never forget how important this geomagnetic field is for life on Earth.

5.7 The Auroral Belt in Olden Times

Where did auroras happen several hundred years ago when the geomagnetic field was much stronger than it is today? According to the International Geomagnetic Reference Field (Alken et al., 2021), the geomagnetic poles have been drifting slowly. Taking many factors into account, it is possible to calculate where the auroral belt was in the past.

Here are some of the factors to consider:

1) Patterns in the magnetic field:

No simple patterns were observed in the increase and decrease of the magnetic field. Moreover, the axis of the geomagnetic field is far from being a perfect dipole, having many local structures. Even if a dipole field could approximate the large-scale magnetic field, its center has shifted from the center of the Earth. When Gauss first measured the intensity of the geomagnetic field, the distance between the two centers was 300 km. Now, it is 500 km.

2) Reversal of the Earth's magnetic field:

This mechanism has not been fully unveiled yet, though we know it has reversed a number of times before. The terrestrial magnetic field must have been extremely weak when the reversal process occurred regardless of the mechanisms. Thus, the reversal story is complicated, to say the least.

Figure 5-8 shows the calculation for the auroral belt's location 300 years ago—it might have been located closer to central Europe, where auroras could have occurred frequently over Scotland, Denmark, Germany and France. To add credence to this theory, scientific research on auroras first began about 300 years ago in Europe, probably in tandem with the auroras that were spotted frequently over Europe at that time. Edmond Halley, the discoverer of Halley's Comet, had even left a record about an aurora being observed in London during the same time period. It was during this era that the romantic-sounding term "aurora" began to be used.

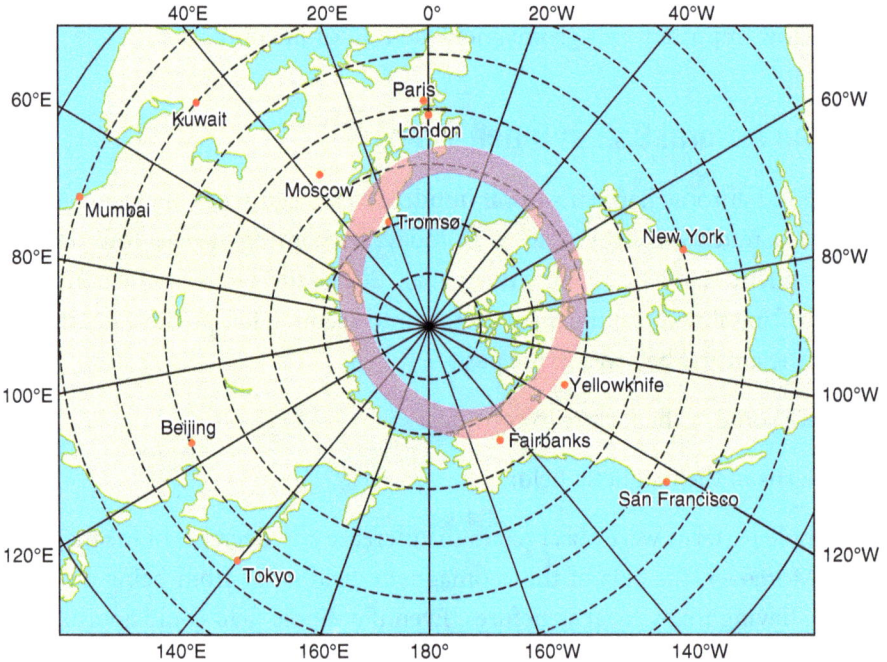

▲ Fig. 5-8. The estimated position of the auroral belt 300 years ago. This estimate was based on the same empirical formula used for the diagram shown in Fig. 5-7. Scotland is found to be well inside this belt. Reconstructed from figure provided by Takashi Oguti.

5.8 Auroras in Old Records

A number of aurora-like descriptions can be found in old documents and have been examined by scientists. Among them, the late Prof. Mitsuo Keimatsu, who majored in Chinese at Kanazawa University, Japan, stands out for his work on auroras. His careful interpretations of old documents, in which aurora-like events are described, are indispensable because people in olden times did not have a good understanding of auroras, and thus were not able to record such events clearly. For example, a description in a Chinese ancient manuscript would read: "A rain of stars fell", prompting some readers to interpret these to be meteorites, while others speculating that it referred to a kind of aurora.

Auroras were commonly expressed as "Sekki", which means "red air" in Japanese. However, we need to be careful in determining whether "Sekki" actually refers to an aurora; it could relate to other phenomena, such as the moon with reddish features as a result of Kosa (yellow sand wind coming from China).

The most ancient record that described an aurora in Japan is perhaps found in the oldest and most authentic document describing Japanese history. Known as "Nihon Shoki", it reports that an aurora was purportedly seen on December 30, in the year 620, during the rule of Emperor Suiko, and was described in the following fashion:

"A red luminary came out in the air. Its height was about 1 Jo with [the] shape of a pheasant's beautiful tail. Subsequently, about 60 years later on September 18, 682, in the reign of Emperor Tenmu, a red luminescence appeared in the northern sky."

It was obviously a description of an aurora.

Teika Fujiwara wrote in "Meigetsuki" about a similar event that occurred in 1204:

"On a clear day, red light came out in the northern sky. It was not a cloud[; it had] 4 or 5 whitish parts and 3 or 4 red lines[—all] moving. It was a horrible scene and scared people".

People in those days thought auroras were harbingers of ill-fated phenomena. Based on some of the known descriptions often used for auroras during that period (e.g., a forest fire, a field fire, a flag, or an unusual shape and color), auroras seem to have "visited" Kyoto several times. An aurora observed on September 17, 1770 must have been very active because many descriptions were recorded. The "Sei Kai" is one such record, and even included a sketch that portrayed the aurora as "red air like a big fire expanding in the sky beyond mountains". It describes: "There seemed to be a big fire in the direction of Wakasa, north of Kyoto, seen from here in Kyoto".

There is another book "Zokushi Gosho" that reports on the same aurora: "People could see the fire from anywhere in Japan. Despite a moonless night, we could recognize each other's faces. A few white lines appeared, and in a short while, it dissipated. The air continued to be red until dawn".

As the 20th century began, a new term, "Hokkyoku-kou" ("Polar Lights"), began to be used to describe the phenomenon. We know this because of an aurora that appeared over Japan on September 25, 1909, that was observed from Akita and Nigata as well as from Hokkaido. The next day, an article appeared in *Otaru Shimbun* (the Otaru newspaper—one of the predecessors of the Hokkaido newspaper, known as the *Hokkaido Shimbun*, or *Doshin*). The article headline (translated) reads as follows: "Kyokkou [Polar Lights] in Hokkaido? A Strange Phenomenon Mystified an Observatory". Two days later, another article, accompanied by a sketch, appeared. It reads: "The light was indeed identified as Kyokkou. An arch of deep-red luminary expanded in all directions in the sky. This rare phenomenon, which occurs 5 or 6 times every 100 years, made people feel uncomfortable". Mr. Sawada, who saw this aurora in northern Hokkaido, said, "I saw light flashing ten times at the left and right. Its color was as red as a plum."

About 1,000 cases of aurora sightings have been collected from archives in China that trace back to a few hundred years before the 1st century B.C.; where auroras are often expressed as "a glittering dragon" (for example, in one record, the aurora sighted was described as "a red animal with a snake body and a human face, measuring as long as 5,000 km." This seems to describe a low-latitude red aurora). Other examples appear to mention active auroras, which usually occur over high latitudes.

5.9 Auroras on Other Planets

Auroras are not unique to the Earth. They can occur on other planets as long as these planets have both air and a magnetic field. Aurora images were captured from the space probe, Voyager I, and the Hubble Space Telescope. In September 1977, NASA had launched Voyager I to study the outer solar system. One and a half years later, it succeeded in taking pictures of an aurora on Jupiter for the first time in mankind's history—the aurora was called a Jovian aurora.

It had long been predicted that auroras exist on Jupiter due to its intense magnetic field, whose intensity is 10,000 times stronger than the Earth's, and strong radio waves are occasionally emitted by the Red Giant.

The locations—latitude 65–75—where Jovian auroras and those on Earth occur at are very similar, but the brightness of the former is 1,000 times more than the latter. It is interesting to also note that the first Jovian aurora was located along magnetic field lines that lead to one of its satellite moons, Io, which is volcanically active.

Following the accomplishments of Voyager I, auroras on Jupiter and Saturn were next filmed from the Hubble Space Telescope (Figs. 5-9 and 5-10). In addition, the presence of auroras on Uranus and Neptune were also established. Although they had different colors (and thus wavelengths), probably due to the different air compositions on these planets, the presence of auroras proves that all four planets have both air and magnetic fields.

◀ Fig. 5-9. Image of Jupiter's aurora taken with ultraviolet wavelengths from the Hubble telescope on December 19, 2000. Credit: NASA, ESA and John T. Clarke.

◀ Fig. 5-10. Images of Saturn's aurora with ultraviolet wavelengths, captured from the Hubble telescope on January 24, 26 and 28, 2004. The images are combined with a visible light image taken on March 22, 2004. Credit: NASA, STScI, ESA, John T. Clarke and Zolt Levay.

Air creates conditions under which life may exist, and it minimizes temperature differences between day and night on a planet's surface. However, the air temperature on the other planets is extremely low because they are very far from the Sun.

Not all planets have a magnetic field. The Earth's core, which consists of fluid metal (melted at the temperature of a few thousand degrees Celsius), can be described as a power plant generating electricity and magnetic fields.

Likely, planets with no air or magnetic fields will not have auroras occurring on them. For this reason, we should not expect to see auroras on Venus and Mars or on the Moon, as they have either very weak or no magnetic fields. The cores of these entities are solid and cannot generate electricity. The planet Mercury is an interesting case. Although it has healthy magnetic fields, no auroras will occur because it is too close to the Sun.

When we view all the above points as a whole, we can only marvel at how life began on our tiny planet called Earth. In a way, auroras are messages to let us know that the air on the Earth and its geomagnetic field protect life from the severe conditions of space.

5.10 Auroras and Intellectual Curiosity

While auroras arouse awe, they also stirred up Romanticism[1] (a movement which was at its peak approximately around the 19th century), during which People with intellectual curiosity started to carefully observe these mysterious lights and began to discover new rules governing auroras.

As aurora science grew in Europe during the 18th to 19th centuries, the science of electromagnetism progressed—natural laws governing electromagnetism were discovered as a result of this growing curiosity. Thanks to electromagnetism, we now have electric appliances that are an indispensable part of modern life. In other words, this life-changing discovery resulted in part from our curiosity about the northern lights.

Through a series of observations and experiments conducted in the 20th century, mankind discovered that auroras are large-scale electricity discharging processes occurring in space (Fig. 5-11). More importantly, while researching the source of power causing these phenomena, the interactive relationship between solar wind and the geomagnetic field was identified.

---

1 "Romanticism", with a capital "R" is not to be confused with "romanticism" (small "r"); the former (as defined in Oxford Languages) refers to "a movement in the arts and literature that originated in the late 18th century, emphasizing inspiration, subjectivity, and the primacy of the individual."

◀Fig. 5-11. Professor Kristian Birke-land (left) carried out a series of Arc-tic expeditions for auroral research in the late 19th century. He is the pio-neer of modern auroral science. He confirmed his theories of auroras by conducting laboratory experiments at Cristenia University, the predecessor of the University of Oslo, Norway. In this picture, Professor Birkeland placed an artificial small Earth with a magnet in the center of a vacuum chamber. The auroral belts were re-produced when plasma flow hit the small Earth. Image courtesy of Asgeir Brekke and Alv Egeland, Nordlyset, page 78, Grandahl, 1979.

This was not the only thing we learned from auroras: they also taught us how all life on Earth is protected from the harsh outer space (where high-energy plasmas from the Sun are constantly buffeting the Earth) through two protective barriers in the form of the air (our atmosphere and ozone layer) and a magnetic field surrounding the Earth. Just like a living object, the Sun is active and breathing and has moods. Auroras are a manifestation of these dynamic changes of the Sun, projected on a giant screen in the air.

Mankind's curiosity has led to further interesting discoveries: the discovery that energy (as much as 3.9×10^{26} Watts) is always emitted from the Sun and that the plasma energy of the solar wind that generates auroras is but only a tiny fraction of the total solar wind energy.

We, humans, deserve credit for our efforts in trying to understand the dynamic existence of the Earth in the universe, with only a few clues to work on. Our sense of curiosity is great, where we attempt to understand every phenomenon in front of us. We are also attempting to go into space in search of the origin of life. Intellectual curiosity distinguishes us from other animals. The simple question "What exists beyond an aurora?" naturally comes from our innate curiosity or perhaps an instinct to seek better understanding.

When we completely understand auroras, we would have also solved the mystery of the solar-terrestrial relationship, which we are currently searching for answers for. An aurora is only one of the manifestations of this delicate balance between the Sun and the Earth.

"As a child, I had a keen appetite for knowledge about nature, and this was the reason I chose to study this scientific field. My quest to uncover the aurora's secret continues with access to data from satellites and computers. I have met a number of like-minded scientists from all over the world, many of whom have had the same desire to know more about nature's mysteries since young."

~ Yosuke Kamide

References

- Alken et al. (2021), International Geomagnetic Reference Field: the thirteenth generation, Earth, Planets and Space, 73, 49, doi:10.1186/s40623-020-01288-x.

- Baker, R. R. (1980), Goal orientation by blindfolded humans after long-distance displacement: possible involvement of a magnetic sense, Science, 210(4469):555-7, doi:10.1126/science.7423208.

- Baker, R. R. (1982), Human navigation and the sixth sense, Simon & Schuster, ISBN:978-0671443900.

- Stoupel, E. G., J. Petrauskiene, R. Kalediene, S. Sauliune, E. Abramson, and T. Shochat (2015), Space weather and human deaths distribution: 25 years' observation (Lithuania, 1989–2013), Journal of Basic and Clinical Physiology and Pharmacology, vol. 26, no. 5, 433-441, doi:10.1515/jbcpp-2014-0125.

- Vanselow K.H. (2020), Where are solar storm-induced whale strandings more likely to occur? International Journal of Astrobiology 19,413–417, doi:10.1017/S1473550420000051.

- Zoeger, J., J. R. Dunn, and M. Fuller (1981), Science, 213, 4510, 892-894, doi:10.1126/science.7256282.

Chapter 6

Expressing Auroras

6.1. What Do Auroras Mean to You?

What comes to your mind when you hear the word "aurora"? Do you imagine auroras to be mysterious, beautiful light displays in the polar sky that hold deep secrets? Or perhaps you might even imagine them to be some super power beyond the reach of human knowledge?

In this chapter, we will see how people's views and superstitions changed with each new discovery of a fundamental characteristic of auroras. We will also look at how artists creatively portrayed their impressions and feelings about auroras.

Auroras can remain dead calm for long periods of time—from several hours to as long as several days. Conversely, very fast changes in both color and shape occur during a substorm. With time as a constant, these fluctuations are visible manifestations of the natural system's response to changes in the solar wind and magnetosphere. See Chapter 4 for the definition and scientific details of substorms.

An aurora is a soundless, monumental and epic presence in the polar sky. Often its form and variability changes dynamically depending on its location in the sky. Its colors also change and may range from whitish green to red, pink, blue and purple. Currently, only those living at the highest latitudes of the Earth can regularly appreciate such beautiful displays of lights. Auroras can also sometimes appear "bloody red" in color, covering the entire sky; thus giving one the impression that something terrible is about to happen. If one were to witness auroral displays without knowing that these lights originated from the Sun, would he or she accept a scientific explanation for this occurrence, or would they choose to believe in the existence of gods?

In any case, every aurora experience is very unique.

I (Kamide) myself experienced one serious event. It was at the beginning of winter on the campus of the University of Alaska, where I was a postdoctoral fellow. One evening, I saw a student walking slowly toward me. The next moment, he suddenly fell down, onto the street about 20 meters away. I immediately rushed over to render assistance. Though he seemed to have serious trouble breathing, he just kept pointing his fingers upward into the sky, blabbering indiscernibly.

After several minutes, his shock and excitement began to dissipate, and he explained what had happened to him. It appeared that he had just arrived the day before from Los Angeles—he had heard of the northern lights but had never seen them. So when he saw lights (or fire) that came from above, he thought that the sky had "broken". He felt nothing else could be done except to pray.

Thanks to the rich progress of Earth and space sciences, we now know that auroras in the polar sky are generated through intricate interactions between the Sun and the Earth, or more precisely, between the solar wind and the Earth's magnetic field. Though many features of auroras have been explained scientifically, people are still moved by the mystery inspired by this phenomenon. Throughout human history, auroras have continuously stirred our feelings and fired our imaginations, thus becoming recurring themes in poems and paintings.

6.2. Fire in the Sky?

Going as far back to the early days of the Roman era, one of the earliest references to an aurora occurrence can be found in 2000 B.C., according to Eather (1980), who had surveyed the historical background of auroras. The following passage in Ezekiel (593 B.C.; Ezekiel 1: 4, 13, 22 [ESV]) can be cited:

"As I looked, behold, a stormy wind came out of the north, and a great cloud, with brightness around it, and fire flashing forth continually, and in the midst of the fire, as it were gleaming metal. […] As for the likeness of the living creatures, their appearance was like burning coals of fire, like the appearance of torches moving to and fro among the living creatures. And the fire was bright, and out of the fire went forth lightning. […] Over the heads of the living creatures there was the likeness of an expanse, shining like awe-inspiring crystal, spread out above their heads."

In Ancient Greece, a record by Xenophanes of Colophon (c. 570–475 BC) described scenes of inflammable exhalations from the Earth and mentioned "moving accumulations of burning clouds" (Stothers, 1979).

Other than forms of fire, in many countries, people relate forms of auroras to various animals. Just as we often think up stories and visualize images while watching clouds in the sky, so too can we do the same for auroras. For example, one can compare the ray structure in aurora lights to swords that people use during wars in ancient times (see Fig. 6-1a for illustration), while a tight aurora core with what appeared like beans of light radiating outward could be seen as a "gigantic snake dancing high in the sky" (see illustration in Fig. 6-1b). Some people may even perceive aurora lights as the result of collisions between a big

▲ Fig. 6-1a. Aurora lights depicted as swords in battles or wars that are going on in the sky. Source: Staatsbibliothek Bamberg.

◀ Fig. 6-1b. Aurora lights as a "radiating giant snake dancing in the sky". Source: The Midnight Cry: Behold, the Bridegroom Cometh by E. M. Hardie, p.453, 1883. See Brekke and Egeland (1983) for more detailed information about this figure.

▲ Fig. 6-1c. In Scandinavia, people once believed that aurora lights were generated by the collision of icy air splashed from a fox's tail and the upper atmosphere. Source: Shutterstock / Simon's passion 4 Travel.

splash of icy snow from a fox's tail and tiny particles in the upper atmosphere.

6.3. Ancient Scholars and Auroras—The Beginning of Aurora Science

Auroras must have appeared in the polar sky even before people began to ask scientific questions about it. In ancient times, philosophers, who were supposed to know the causes of all natural phenomena, spoke of auroras by invoking spirituality and the gods. It is widely known that the Greek philosopher Aristotle (384–322 B.C.) tried to explain natural phenomena in terms of four basic elements: fire, air, water and earth. According to his theory, auroras were simply the result of collisions between vapor (air) after being heated by the Sun (fire). We now know, of course, that this scenario is not possible.

The Roman statesman Seneca the Younger (4 B.C.–65 A.D.), who was a philosopher, noted the following after the time of Aristotle: "There are *Chasmata* (chasms), too, when there is a subsidence of some portion of the heavens, which sends out hissing flame, as it were, from its hidden recesses. There are also a great number of colours in all these. Some are of brightest red, some of light insubstantial flame, some of white light, some glittering, some with a uniform glow of orange without sparks or rays." (Quaestiones naturales, Liber I, 14.1-14.2, trans. Clarke, 1910.)

It is interesting to see how many people at that time regarded auroras as something superstitious, as evidenced by Seneca's attempt in describing what he saw in Rome. The aurora that he saw was actually a low-latitude aurora that occurs under extremely high solar activity.

For some reason, the great Renaissance did not contribute very much toward the understanding of what auroras were. Still, this period of history cannot be complete without mentioning the two great astronomers of that time: Johannes Kepler and Galileo Galilei.

Both astronomers discovered a number of basic laws about planets and the Universe that control virtually all natural phenomena in the Universe as well as on Earth. Kepler was finalizing his theory about the motions of the planets, while Galileo was organizing his observations by using his first telescopes. They both left partially correct descriptions of the auroras they saw in Italy.

Several more scientists who contributed to progress in the study of the generation mechanisms of auroras are:

- Pierre Gassendi (1592–1655), a French mathematician as well as an astronomer. He thought that the vapor of auroras was associated with the production of the lights at high altitudes.

- The great philosopher René Descartes (1596–1650) hypothesized that auroras were nothing but a scattering of sunlight from ice particles in the cold polar sky.

- Edmond Halley (1656–1742), an astronomer, studied the corona-type aurora in detail, where rays of light scattered radially. We now know that this type of aurora is a signature of a substorm's breakup.

- Jean-Jacques d'Ortous de Mairan (1678–1771), a French geophysicist, published his theory of auroras in 1731. He insisted that auroras were a reflection of sunlight from polar ice and snow in the polar atmosphere. Though it should be noted that de Mairan's book was widely read, this did not mean that his ideas were well accepted. For example, mathematician Lenhard Euler (1707–1783) assumed that auroras came from particles in the Earth's atmosphere.

- John Dalton (1766–1844), an English chemist, calculated the height of auroras by using the so-called triangular method. This same method was conducted extensively later on by the Norwegian astronomer Carl Størmer (1874–1957) (see Sec. 4.3) as well. Dalton described the beauty of auroras as follows:

> "The whole hemisphere was covered with [auroras] and exhibited such an appearance as surpasses all description. The intensity of the light, the prodigious number and volatility of the beams, the grand intermixture of all the prismatic colors in their utmost splendor, variegating the glowing canopy with the most luxuriant and enchanting scenery, affording an awful, but at the same time, the most pleasing spectacle in nature."

6.4. Origin of the Term "Aurora"

"Aurora" originated from Roman mythology, and originally referred to the Goddess of Dawn who would bring the sunrise to the Earth every morning, so that all creatures on it could begin their day with a beautiful dawn (see Fig. 6-2). Ancient people believed that owing to her great power, the world would have

▲ Fig. 6-2. The Roman goddess Aurora. The function of this god was to chase away all the stars in the dark sky and bring in new stars for the next day. Credit: Adobe Stock / Archivist.

a systematic and vivid time system. Without her, a dark Earth or a hopeless world would endure forever.

It was only after the 17th century that people began referring to the lights in the polar sky as auroras. Who was responsible for this name? The scientific community seems to agree that it was Galileo who began to use this beautiful word after seeing a spectacular show of lights in northern Italy. If this is indeed the case, we can say that the same person discovered both cause and effect, i.e., sunspots on the Sun and auroras on Earth.

6.5. Exploration of Polar Routes and Auroras

Observations and descriptions of auroras after the 19th century cannot be made without mentioning the brave men who faced challenges in what is now called the last frontier. In particular, polar explorers from Europe to the Orient through northern Canada aimed to reach the North Pole. Others tried to open polar routes for future commercial business by using ships. Because of bitter and severe weather conditions in the polar region, many teams gave up pursuing their adventures. Many of these expedition parties never returned to Europe. These explorers may have witnessed auroras from their boats, albeit in mental despair, as they did not know whether they would return from their voyages.

Perhaps one of the most famous and dramatic explorations was made by Fridtjof Nansen (1861–1930) of Norway. He kept a diary chronicling his extensive exploration of trying to open a new polar route. Nansen was famous not only for his great exploration voyages but for also being a scientist, statesman, and artist. Perhaps the following description by one of his crew members during one of his winter expeditions will shed some light on how much he was moved by this splendid natural show in such a difficult time:

"Shortly after midnight, Nansen left us for a quiet stroll across the ice. It was a beautiful clear night, with the streamers of the aurora borealis shifting across the heavens. He turned and looked back to see the dark masts against the yellow glow of sky. Behind it, the silken draperies of light were shimmering across the heavens like great pulsating rays of violet sheen, intermingled with pastel shades of pink and green. For almost an *hour, he stood there. In the cold silence, he thought of the long months they had already spent in this vast frozen wasteland and even longer months that might lie ahead. He thought of Norway and of home for a fleeting moment.*" (See Fig. 6-3.)

6.6. Folklore and Legends

Auroras include a number of associated processes and events. They also hint at folklore, depending on which processes people felt were important

◀ ▲ Fig. 6-3. Even under heavy pressure, bearing great responsibility and living in desperate conditions, crew members of F. Nansen's team were always encouraged to observe the sky. The person depicted standing on the icy field beside his boat must have been Nansen himself. Courtesy of Tromsø Geophysical Observatory.

and essential to them. For example, ancient Chinese names for auroras were "candle dragon" and "cracks in heaven", while the Aboriginal North Americans and Inuit described them as "fun ball plays in the sky by people who died on the ground"—some believe auroras to be a form of communication between themselves and the spirits of those who recently passed on. In other words, these beams of light connected heaven to earth.

According to a widely held Inuit legend, a just-departed spirit could enter the body of a newborn. Some also believe that departed spirits can go on to various levels of existence, depending on a person's behavior in life. The highest level was the aurora's place in heaven—a happy place where there are no stormy days. Another legend has it that auroras are the spirits of their ancestors playing a game of football.

The origin of Fig. 6-4 is uncertain, but it tells of how the people in Europe commonly believed that auroras were "gigantic candles in the sky". It is interesting to note that stars are located well below auroras in their descriptions.

In Scandinavia, auroras were believed to be a "bridge" for dead people to return to the Earth. People believed this to be true because auroras "responded quickly" whenever they were waved.

▲ Fig. 6-4. Once, people believed that "many candles in the sky" were the cause of auroral lights. Source: Royal Observatory, Edinburgh.

6.7. Auroras in Art

Artists who witnessed auroras are usually so moved by their natural beauty that they feel compelled to bestow what they felt and imagined at that time unto others through their creations.

An example of such an artist is Harald V. J. Moltke (1871–1960), a Danish artist who is perhaps the most historically well-known painter of auroras. He traveled to Scandinavia and Iceland many times, leaving impressions of the auroras he witnessed in the form of stone cuttings and oil paintings. His paintings captured auroras from the perspective of a photographer looking up at the natural lightshow, and typically include a person watching the auroras.

This technique made the auroras in his work look natural because of the many small-scale structures in there. When I (Kamide) saw his works for the first time, I stood frozen, unable to move for about 10 minutes or so. While visiting the Danish Meteorological Institute, I found his work displayed at the entrance hall by the Institute's library. I was surprised to see that the color and various shapes in his work were very realistic. Truly, it made for a grand entrance. Figure 6-5 is one such painting by Moltke.

◀ Fig. 6-5. A painting by Harald V. J. Moltke.

Let us introduce two presently active professional artists who paint auroras—one is a Canadian who sees auroras almost every night, while the other is Japanese. It is quite interesting to see how both artists added their own perspectives to what they saw.

Robbie Craig (see Fig. 6-6a) is a professional artist working out of Yellowknife, Canada. Originally from Barrie, Ontario, Craig's adventurous nature led him to the Northwest Territories in 2006. Instantly inspired by the beauty of Canada's North, he connected with his creative side and rediscovered his childhood passion for art. His work embodies the natural rugged beauty of the landscape that surrounds him. With an impressionistic style that is truly his own, he creates colorful pieces that have a unique feel. Here are some questions he answered for us. Figures 6-6b through 6-6d show three of his paintings.

◀ Fig. 6-6a. Robbie Craig: An artist living in Yellowknife, Canada.

▲ Fig. 6-6b. 'Echoes in the Distance' by Robbie Craig.

▼ Fig. 6-6c. 'Bison under the Aurora' by Robbie Craig.

▼ Fig. 6-6d. 'Bison under the Aurora' by Robbie Craig.

What caused you to become interested in arctic life and auroras?

"One of the reasons I feel so fortunate has to do with my place; the beautiful landscape of northern Canada. A place where I can view the majestic northern lights from my bedroom window and drive only a few short miles to feel a million miles away. I have found and made my home in Canada's Northwest Territories. A place where true adventure still exists; from driving across Great Slake Lake on ice roads to chase the aurora amid −40°C weather, to the contrasting warm days of 24 hours of daylight in the summer, the north holds unique experiences that you cannot find anywhere else. The north feeds my soul and inspires my work."

What are auroras to you?

"Each time I view an aurora, I feel like it is my first. To me, it is a reminder of the power that exists within the universe, a reminder of something greater than ourselves. Each time I saw an aurora, I realize that my heart deep in my body gets washed out by a big power existing in the natural system. Although I understand the science behind this natural phenomenon, to me, the aurora is something ethereal, something magical."

Website: www.rcraig.org

Tokio Nakatani, who represents the senior generation, lives in Japan. Born in Chiba City near Tokyo, he has been very active in expressing natural phenomena. Since he cannot see auroras in person every day, his product is the result of competing motions between his imagination and real auroras (see Fig. 6-7).

He invented a method, which is quite useful to express messages from auroras, called the Dividing Space Method. Realizing the difference between what he imagines and what he sees in reality, he confesses that challenging auroras is like challenging life. He has received a number of medals and awards for his excellent works. Figure 6-7 shows one of his recent works.

You can find out more about him on his website (in Japanese): http://www.shinkozo.or.jp/photography_course.html

▲ Fig. 6-7. A product of the multiple arc system by Tokio Nakatani, a Japanese artist.

6.8. Auroras in Music and Poems

Theodor Caspari (1853–1948) is a Norwegian poet from the 19th century. As one of the examples of Caspari's expressions of auroras, part of the poem "Northern Light" is shown below:

Northern Light

Sparkling Arc,

playing Flame.

Flickering garland around the mistful Pole.

Iceful blaze

pictureless Frame

you are to me, Aurora, a symbol of life.

The thoughts are created,

strengthened and scattered,

twined and twisted in Generations by Hands.

The sparks are meeting,

agreeing and linking

Generation after Generation as flickering Bands. –

Twinkling stars,

rolling Spheres

circling aloft in light Harmony.

Thoughtful Brains

Modest Heads

Ignite down here a Firework.

See, how it is shining

Lightning Lances,

Silver Tiara of the Illumination's Treasure

The Generations are freezing

rejoicing and dancing

Restless about in the eternal Night. –

Caspari also tried to explain what aurora light was by using descriptions like "ice needles". It is quite interesting to note that Willy Stoffregen (1909–1987), a Norwegian physicist who worked on a theory of radio propagation associated with aurora brightening, composed music based on this poem. Stoffregen had been passionate about both the sciences and the arts.

Among many other poems, there is a famous one where auroras are the main subject. It was titled "Autumn" by Scottish poet James Thomson (1700–1748), from which the following stanza is taken:

> Oft in this season, silent from the north
>
> A blaze of meteors shoots, ensweeping first
>
> The lower skies, then all at once converge
>
> High to the crown of heaven, and all at once
>
> Relapsing quick, as quickly reascend,
>
> And mix, and thwart, extinguish, and renew,
>
> All ether coursing in a maze of light.

Other poems like "Driftwood" by Henry Wadsworth (1807–1882) among many others by other poets, including Goethe (1749–1832), Byron (1788–1824), Keats (1795–1821) and Emily Dickinson (1830–1886), refer to auroras in their poetry.

It is clear that there is an intricate relationship between art and science when it comes to our understanding of the natural world.

References

- Brekke, A. and A. Egeland (1983), The Northern Light: From Mythology to Space Research, Springer, ISBN 978-3642691089.

- Clarke, J. (1910), Physical science in the time of Nero being a translation of the Quaestiones naturales of Seneca, p.38, doi:10.5962/bhl.title.1912.

- Eather, R. H. (1980), Majestic Lights: the aurora in science, history and the arts, Am. Geophys. Union, ISBN 978-1-118-66496-4.

- Stothers, R. (1979), Ancient Aurorae, ISIS, 70, pp.85-95, doi:10.1086/352156.

Appendix

Getting the Best Out of Your Aurora Viewing Experience

As demonstrated in this book, the aurora is a unique manifestation that shows us how the Sun interacts with the Earth, the conditions of solar wind, and how much our planet is receiving energy from the dynamically-changing Sun. Auroras are not only beautiful light shows in the polar sky; they provide us with hints on how our planet is protected from the harmful materials outside our atmosphere.

A.1. Where Can We Go to View Auroras?

The most popular destinations for aurora viewing are locations encircling the geomagnetic poles. This is called the auroral belt or the auroral zone. The Earth has two belts of lights, one in the northern hemisphere and the other one in the southern hemisphere (Fig. 1-1). In the chapters of this book, the pictures of auroras taken in the northern and southern hemispheres are shown.

For auroral tourists, the auroral belt or auroral zone in the northern hemisphere, rather than in the southern hemisphere, is the preferred destination, because more commercial flights are regularly scheduled to get there.

So where should we go to view auroras? Although auroras can be seen at different locations along the northern auroral belt they do not differ considerably. The following are the locations most frequented by auroral tourists.

Alaska

The entire state of Alaska, from Barrow in the north to the south of Anchorage, is embedded within the northern auroral belt. Fairbanks, located near the center of the state, has the highest probability of seeing intense auroras and is home to the University of Alaska, where they have an active aurora research

▲ Fig. A-1. Example of an auroral breakup, which signals the beginning of an auroral substorm. People say that if you happen to be just below the breakup, you feel like you are in the middle of shower of lights. When I'm asked "What is the aurora to you?" one of my answers would be "It provides me with hints toward understanding solar–terrestrial relationships" and "It shows the most beautiful natural phenomena that we see on the planet Earth." Photo by Y. Otsuka.

program. Fairbanks is easy to reach by air and has many amenities for visitors. It is cold in winter, reaching as low as -40°C, but has relatively low humidity to make life there not so difficult.

Canada

The auroral belt in Canada runs from the Alaska-Canada border in the Yukon to the east coast of Hudson Bay. For ease of travel by air, three northern destinations are accessible for aurora viewing in Canada. One is Whitehorse in the Yukon, the second is Yellowknife Northwest Territories (NWT), and the third is Churchill, Manitoba at the western edge of Hudson Bay. Auroras do sometimes visit Ottawa, the capital of Canada, once or twice a month depending on solar activity.

Yellowknife is perhaps one of the few most popular destinations for aurora tourists in the world. It is the capital of NWT, with a population of approximately 20,000 people. A statistical study indicates that a three-night stay in Yellowknife has a 95% probability of encountering auroras for at least one night. This rather high probability earns Yellowknife its nickname, Aurora Ginza*, by Japanese visitors.

*Ginza = a worldwide known shopping district in the center of Tokyo.

Scandinavia

Three countries share northern Scandinavia, where people can enjoy auroral displays whenever the weather is favorable. They are: Norway, Sweden and Finland. To the east of these Scandinavian countries is Murmansk in Russia, which attracts many visitors as well. It is home to a research institute specializing in geophysics and space science.

Between the eastern edge of Canada's North and the three Scandinavian countries, two additional places attract aurora tourists: Iceland and the southern edge of Greenland. However, they are not as popular as Canada and Scandinavia because of the long flights to get there.

All in all, those places located along the auroral belt attract a number of tourists from the countries around the world where auroras are not normally seen. Again, auroras seen at different places are not very different in appearance. Depending on the longitude of the location, you can enjoy various types of auroras in different landscapes or settings. For example, the corona type of auroras seen in a Norwegian forest versus those breakup auroras reflected on the surface of lakes near Yellowknife provides very different photographic impressions.

The authors of this book have seen more than 1,000 auroral breakups in many different places in the world. Of course, each of them was so beautiful, but no two were alike or linked. Figure A-1 shows one that was occurring just above their heads, while Fig. A-2 shows a more large-scale signature of an auroral breakup. As discussed in Chapter 4, perhaps the most beautiful, as well as dynamic auroras, relate to auroral breakup and its byproduct.

A.2. At What Time Do Auroras First Appear?

If you are already at auroral latitudes, your chances of encountering auroras are good whenever the sky is dark. However, auroras are not a random process; they do follow physical laws, showing statistically at a peak hour. For example, if a geomagnetic storm is in progress, you can expect auroral displays at any time after 1800 hours Local Time (LT). If an intense geomagnetic storm is in progress, caused by high solar activity or by high-efficiency coupling between the solar wind and the Earth's magnetic field orientation, the probability of seeing nice auroras is quite high. Under such favorable conditions, you can expect to see bright, fast-moving auroras even before sunset, and this high probability of an auroral appearance would continue into the night.

According to statistics, the time with the highest probability in which the brightest aurora appears is between 2300 hours and local midnight in Magnetic Local Time (MLT). Note that midnight in MLT differs considerably from LT, which we use for daily life, depending on longitude. How much those two times differ depends on the Universal Time (UT) and the season. A simple assumption leads us to the following table, showing roughly LT in MLT.

▲ Fig. A-2. An example of a typical westward traveling surge (WTS) in the pre-midnight sector that results from auroral breakup in the midnight sector. No two breakup auroras are alike. In cases of the onset of multiple substorms, the number of WTSs represents the number of auroral intensifications.

Alaska – Fairbanks   0213 LT

     – Barrow   0306

Canada – Yellowknife 0110

     – Churchill   0039

Greenland – Narssarssuaq 2241

Iceland – Leirvogur   2359

Norway – Tromso   2239

Sweden – Kiruna   2243

Finland – Lovamiemi 2325

Russia – Murmansk 2302

It is immediately noticed that maximum auroral activity is after midnight in Alaska and northern Canada, whereas it is before midnight in Europe.

Look to the northern horizon for the first appearance of an aurora

People, even tour guides, often say, "We have no idea where an aurora first appears. Thus, we must watch the sky very carefully as we do not wish to miss any minor light, particularly the first one of the day." This is not true at all.

The first aurora of the evening always comes from near the northern horizon.

In this context, the north is the geomagnetic north. The separation angle between geomagnetic north and geographic north depends on the longitude of your location.

Alaska – Fairbanks   +28 degrees

     – Barrow   +26

Canada – Yellowknife +27

     – Churchill   +2

Greenland – Narssarssque -33

Iceland – Leirvogur   -20

Norway – Tromso   +7

Sweden – Kiruna   +6

Finland – Lovamiemi +8

Russia – Murmansk +12

To help determine the direction of the geomagnetic north, you can use a compass for your location. Auroras come from the geomagnetic north every day. It would be a good idea to try to identify the direction of the magnetic north every evening when you are out aurora viewing. If you find a slight indication of a faint light near the northern horizon, you have a chance of seeing gorgeous aurora activity before your night is over.

A.3. How Does an Auroral Breakup Begin in Real Time?

As has already been explained, an aurora substorm begins with a special type of aurora show that appears right at the beginning of a magnetospheric substorm. A substorm has three distinct phases: a growth phase, an expansion phase, and a recovery phase. It is agreed that during the growth phase, energy from the solar wind is extracted and is being stored in the magnetotail. During this phase, some energy is leaking away from the plasma sheet of the magnetotail and generating a diffuse aurora. The beginning of the expansion and recovery phases begins with the appearance of intense and rapidly moving discrete auroras, signaling sudden dumping of particle energy onto the polar ionosphere. It is common for this breakup aurora to be bright and moving rapidly.

People often say that they are at a loss for words on how to properly describe the beauty of an aurora and its variations, which dance before their eyes. This is especially the case when they have just witnessed an auroral breakup. The question is, are we able to predict when a breakup occurs?

In Fig. A-3, an example of an auroral breakup is seen in three colors, plotted into a latitude UT coordinate system. The time shown at the bottom is UT. In this particular case, breakup occurred three times continuously, 2040 UT, 2103 UT and 2130 UT, as shown by a triangle. A rather sharp poleward movement of the aurora characterizes the breakup.

The important key is to find the equatorward expansion of the auroral oval that takes about 30–60 minutes, which leads to an occurrence of an auroral breakup. The equatorward expansion of the auroral oval in the midnight sector means that the size of the polar cap is expanding to accommodate the increasing energy. The idea is that as long as the energy in the polar cap is being stored, the energy is to be released soon. That is the breakup.

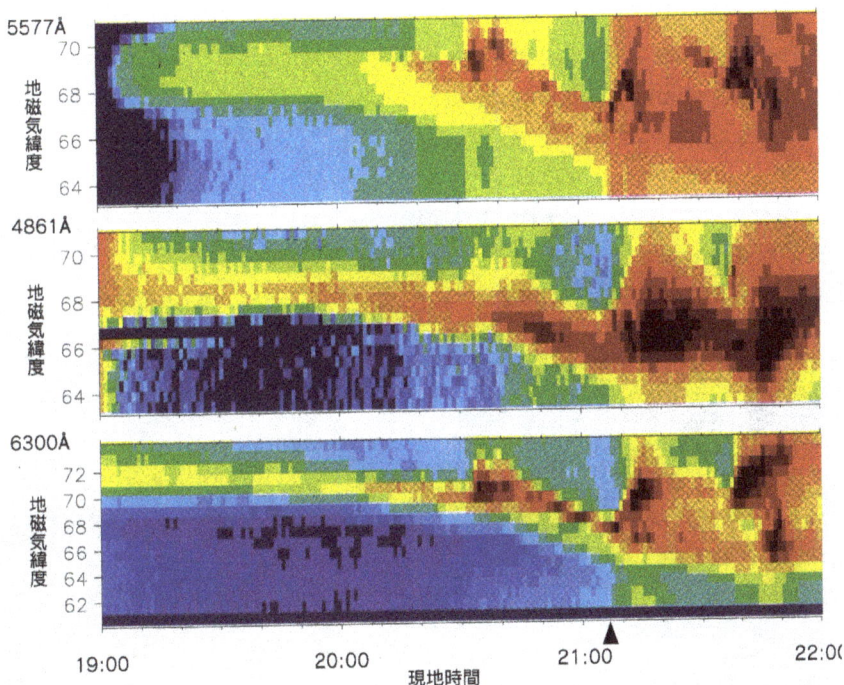

▲ Fig. A-3. Auroral behavior prior to auroral breakup (indicated by a triangle in the x-axis) in three different colors: (from top) green, purple, and red. All three plots are provided in "geomagnetic latitude" (y-axis) and time (x-axis, in Universal Time). Image courtesy of Gordon Rostoker.

A.4. Can You View Auroras from an Airplane Window?

It is often the case that during a long international flight, you can encounter bright auroras. If you take an evening flight from Tokyo to New York, you may be able to encounter an after-dinner show on the left-hand side of the aircraft. If you are very lucky, you may enjoy viewing auroras several times before sunrise the next morning.

This is a similar experience on flights from Europe to Asia. Figure A-4 was taken by Yasuo Takeda, a close friend of the authors. This was taken just before the landing of a Tokyo–Seattle flight, showing a high-energy aurora in purple along field lines (a sign of sunrise), together with the wing of the airplane. When you view an aurora from a plane, the aurora appears to be at nearly the same altitude as an airplane. This impression, however, is an optical illusion.

It is not very surprising to feel that one is flying at nearly the same height as an aurora. However, you have to take into consideration the curvature of the Earth and the reality of auroras occurring hundreds of kilometers from the ground as explained in Chapter 4.

▲ Fig. A-4. An aurora seen through the windows of airplane. This particular photo was taken by Yasuo Takeda on the way from Tokyo to Seattle.

A.5. Can We Hear Auroras?

Some people have wondered if auroras produce sound, and if so, can we hear it? This is one of the long-existing questions among aurora circles in the world. In fact, a number of people have claimed that they heard certain sounds from auroras during peak activity. According to some systematic studies in Finland, a number of people surveyed claimed that they have heard a very special sound, like a hiss or a whistle coming from the aurora.

It is still a mystery, however, as nobody has been successful in recording the sound associated with an aurora. Therefore, any phenomenon that is reported anecdotally deserves a degree of skepticism. Only when solid evidence is available in the form of, for example, repeated instrumental recordings, are people likely to be convinced of auroral sound. Few such attempts have been made in this regard.

Sound is indeed an acoustic wave, which must traverse a medium, like air. The traveling speed of sound is determined by the temperature of the medium. We know that the speed of acoustic waves in the air is 320 m/sec when the temperature is 20°C. This means that it takes about 20–30 minutes for sound waves to travel in the air between an aurora's location and the listener on the ground. People who claim that they heard sounds from auroras report that when visible auroras move, the sound also changes simultaneously. Given the 30 minutes it would take for any sound to travel from an aurora to an observer on the ground, the only possible mechanism for any sound to reach the Earth's surface at a faster speed is if that auroral sound travels as electromagnetic waves and are then somehow changed into acoustic waves. If this were the case, it would mean that those who can hear the auroral sounds have superpowers that can change the form of the waves in the atmosphere!

A.6. Predicting Auroral Activity in Real Time

Real-time information about auroral appearances is now available. Access to this information is not only limited to auroral researchers but can also be used by anybody interested in this phenomenon to predict auroral activity.

To predict world auroral activity, it is easy to extract the "present" values of key variables from some of the observatories and research institutes that make useful aurora data available through their websites:

- Swedish Institute of Space Physics (http://irf.se//Observatory/?link=All-sky_sp_camera)

- Kjell Henricksen Observatory, UNIS (http://kho.unis.no/kho_sony.htm)

- Yellowknife, University of Calgary (https://www.asc-csa.gc.ca/eng/astronomy/northern-lights/auroramax-observatory.asp)

- Mr. Yuzo Koga, Live! Aurora c/o Pokar Flat, University of Alaska (http://aulive.net/openlive/)

- National Oceanic and Atmospheric Administration (NOAA), Space Weather Prediction Center (http://www.swpc.noaa.gov/)

It is fun to check these websites for the present status of the solar wind, or if you would like to see changes in the interplanetary magnetic field (IMF). For instance, it is known that if the IMF is directed southward, auroral activity becomes active. More specifically, when the IMF turns from northward to southward, the magnetosphere begins to store energy for substorms. Under these conditions, the result is a high probability of a nice aurora breakup. On the contrary, if the IMF turns northward, it is a sign that the aurora will become "quiet".

www.ingramcontent.com/pod-product-compliance
Lightning Source LLC
Chambersburg PA
CBHW050630190326
41458CB00008B/2213